2nd Fully Revised and Enlarged Edition

Extension Education, Management in Veterinary Science & Animal Husbandry

NIPA® GENX ELECTRONIC RESOURCES & SOLUTIONS P. LTD.
New Delhi-110 034

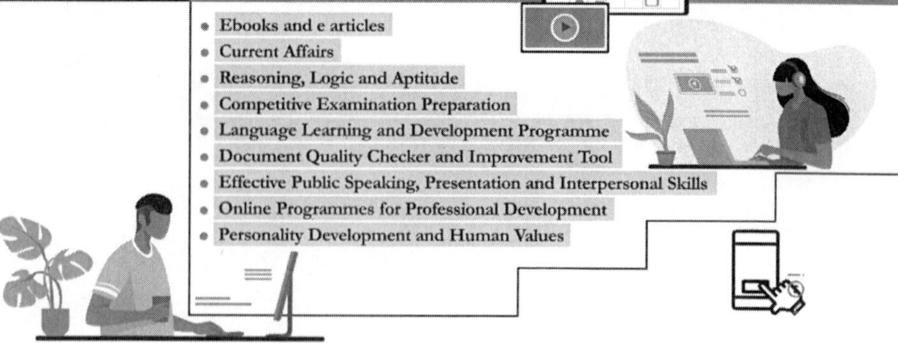

ONLINE RESOURCES
Ebooks and e articles
Current Affairs
Reasoning, Logic and Aptitude
Competitive Examination Preparation
Language Learning and Development Programme
Document Quality Checker and Improvement Tool
Effective Public Speaking, Presentation and Interpersonal Skills
Online Programmes for Professional Development
Personality Development and Human Values

2nd Fully Revised and Enlarged Edition

Extension Education, Management in Veterinary Science & Animal Husbandry

Ruchi Singh Ph.D.
Assistant Professor
Veterinary & Animal Husbandry Extension Education
Nanaji Deshmukh Veterinary Science University
Jabalpur, Madhya Pradesh

M.K.Mandal Ph.D.
Professor
Veterinary & Animal Husbandry Extension Education
West Bengal University of Animal & Fishery Sciences
Mohanpur, Nadia, West Bengal

NIPA® GENX ELECTRONIC RESOURCES & SOLUTIONS P. LTD.
New Delhi-110 034

NIPA® GENX ELECTRONIC RESOURCES & SOLUTIONS P. LTD.

101,103, Vikas Surya Plaza, CU Block
L.S.C.Market, Pitam Pura, New Delhi-110 034
Ph : +91 11 27341616, 27341717, 27341718
E-mail:newindiapublishingagency@gmail.com
www: www.nipabooks.com

For customer assistance, please contact
Phone: + 91-11-27 34 17 17
Fax: + 91-11- 27 34 16 16

ISBN: 978-93-95319-91-1

Composed and Designed by NIPA.

Preface to the Second Edition

Veterinary and animal husbandry extension and management as a discipline and profession in all dimensions. The most important aspects of its always on socio economic transformation of rural community. It also combined different stakeholders who directly or indirectly contribute to the success of farming as well as better livelihood of all concerned rural households.

In tune with the advancement of veterinary science, it is equally important to take forward the innovations from research institute to the farmer's field. Hence, for effective and efficient technology transfer, trained manpower is necessary who can channelize need and location based research to farming community. Students who are ambitious and aspiring to become experts in the field of extension education, this book compile the theoretical as well as multiple objective questions of whole syllabus of JRF, SRF, ARS, NET etc. competitive examination.

The second version of book, titled 'Extension Education, Management in Veterinary Science and Animal Husbandry', has been rewritten and enlarged. The first edition proved to be quite useful for student preparing for competitive examination in extension education. According to feedback and suggestions from students, researcher and professional, efforts are being made to improve the contents quality.

The authors greatly acknowledge all the sources of references from where this book has been compiled.

Authors

Preface to the First Edition

Presently very few numbers of books are available on Extension Education in Veterinary and Management and also to the best of my knowledge; there is no book which deals theoretical as well as competitive aspects of Extension and Management in Veterinary and A.H. Further with new pattern of competitive examination, the need for book which containing theoretical and multiple choice questions both together are aroused.

This book is proposed to make a beginning in this direction by briefly discuss the theoretical aspects of Extension Education in Veterinary and management in ten chapters i.e. Extension Education and Entrepreneurship, Women Empowerment, Rural Sociology, Research Methodology in Extension, Communication Process & Transfer of Technology, Extension Teaching Methods, Programmes for Rural Development, Management in Extension, Information Communication Technology and Statistical tools in Extension Research with each chapters having multiple choice questions.

During compilation of this book and its multiple choice question we have consulted various papers, test books, references book and internet sites from which so much guidance got it. So we are very thankful to all the esteemed scientists and authors whose book, bulletins, articles and sites we have consulted in the preparation of this book. The all sources of information are acknowledged at the end of this book in the form of references.

With the blessing of gods, my family, my teachers and advisor, smile of my daughter and the timely help of so many persons, we were able to complete the book in time.

We also want to express my gratitude to my publishers who have made every effort to present the book in time.

I am indeed grateful to my husband Mr. Nishant Singh and my lovely little daughter Srishti for cooperating during the period of writing this book.

We feel that this book will serve as a competitive book for student preparing for PhD admission test, NET, SLET, JRF, SRF, IRMA, ARS and civil services examination in concerned subject. As this book help them to revise the entire course with objectives questions in a shorter time period.

Any suggestions for further improvement in the book, from the academic fraternity and the students are most welcome. They would be taken note of and incorporated in next edition.

Authors

Contents

1

Extension Education (Theory)

- **Father of Extension:** James Stuart
- Word Extension: Derived from Latin roots "ex" - Out
  "tensio" - Stretching
- Extension term originated in England as University Extension in 1866.
- The term Extension Education first used in 1873 by Cambridge University.
- Philosophy of Extension (Kelsey and Hearne, 1967): How to think.
- Binet test: Intelligence test (IQ)
- **Function of Extension:** Change in Attitude, Skill and Knowledge of an individuals.
- **Extension Education Process:** Extension education is a participative and continuous process
- **Learning Situation:** It is a condition or environment in which all the elements necessary for promoting learning are present.
- Elements of effective learning situations in Extension are:

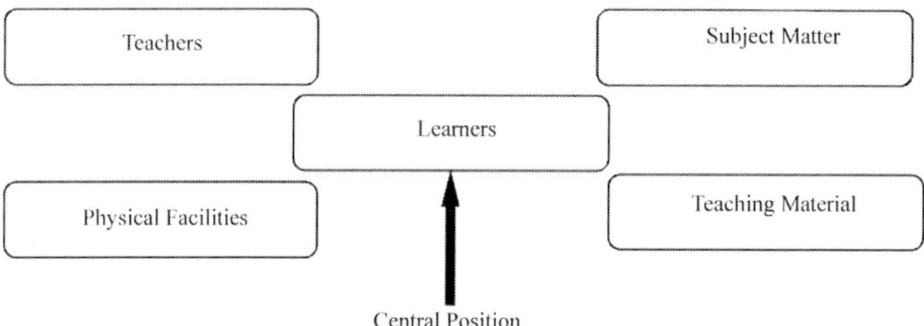

Motivation: Defined as the incentive that is given for inspiration to accomplish something. Motivation can persuade someone to think a certain way or a drive that moves them to do a certain thing.

1. Intrinsic Motivation: Within one's own self eg: Good work for satisfaction
2. Extrinsic Motivation: Generate from outside activity eg: Certificate, Medal etc.

Theory of Motivation

1. **Expectancy Theory:** According to this theory employees are motivated if an employee does very well and puts additional effort, they will likely expect to be rewarded accordingly. If this expectation is not met, employees become unmotivated.
2. **Equity Theory:** According to this theory employees are motivated if they feel that they are being treated equally. If two employees perform the same job, and believe that they do so equally well, they would expect equal pay and equal recognition. Lack of equity, either real or imagined, can damage the employee motivation.
3. **Maslow's Hierarchy of Needs:** According to Maslow several levels of needs that people seek to fulfill. He described in a form of pyramid, these levels are:
   1. Physiological (basic need for a place to live and food to eat).
   2. Safety (desire for security).
   3. Belonging (want to feel like a part of something).
   4. Esteem (the need for praise or respect).
   5. Self-Actualization (the fulfillment of one's potential).

Type of Education

Formal Education	Informal Education	Non-Formal Education
Planned with a particular end in view.	Incidental & spontaneous Not-pre-planned.	Derived from the expression 'formal education.
Limited to a specific period.	Not imparted by any specialised agency.	Outside the realm of formal education.
Well-defined and systematic curriculum, Observes strict discipline.	No prescribed time-table or curriculum.	Conscious and deliberate. Serving the need of the identified group
Given by specially qualified teachers.	Includes activities outside the classroom May be negative also	To be organised for a homogeneous group.

Extension started by

In	Japan	:	Imperial Agricultural Association
	USA	:	Cooperative Extension & Home Economics services
	Israel	:	Government and Jewish agency of Settlement Department

Taiwan : Farmer's Association

China : National Committee of Chinese YMCA in 1915

Name of Extension education / programmes in different countries

1.	Indonesia	Field School
2.	Philippines	Massagana 99 / School on the air
3.	China	Harvest Programme
4.	USA	National Grange
5.	Isreal	Kibbutz
6.	Latin America	Radiophnics
7.	Nepal	Tukki Approach
8.	Brazil	Cruzada ABC
9.	Canada	Media forms

Extension education

According to J.P. Leagans (1961) "Extension education is an applied science consisting of content derived from research, accumulated field experiences and relevant principles drawn from the behavioral science synthesized with useful technology into a body of philosophy, principles, content and methods focused on the problems of out of school education for adults and youth."

According to Kelsey and Harne (1963) "Extension work is an out of school system of education in which adult and young people learn by doing. It is partnership between the Govt. and the people, which provides service and education designed to meet the people. Its fundamental objective is the development of the people."

According to O.P. Dahama (1973) "Extension Education is defined as an educational process to provide knowledge to the rural people about the improved practices in a convincing manner and help them to take decision within their specific local conditions."

Concept of Extension Educational Process: Extension is a continuous educational process, in an effective extension education programme, the following five stages are included (Leagans, 1971):

1. Study of situation and problem– What it is now
2. Objectives and solution– What should be
3. Teaching plan of work– What to teach and how to teach

4. Evaluation– What has been the result
5. Reconsideration– What next to do

Steps in Extension Teaching

Extension teaching is a planned and deliberate act on the part of the extension worker in which extension worker act as a facilitator and motivator. Though procedure may vary from situation to situation. There are some steps which are basic to extension teaching. These are presented following by Wilson and Gallup (1955) (AIDCAS):

					SATISFACTION
				ACTION	
			CONVICTION		
		DESIRE			
	INTEREST				
ATTENTION					

Levels of Extension: Extension is generally thought of at two levels.

1. Extension education: In this level of extension education role is performed by higher learning institutions such as Agricultural, Veterinary universities and colleges, ICAR institutions and Training and extension organizations
2. Extension service: In this level of extension, provide educational service to the people according to their requirement. By this service peoples improve their life through better working. The service provide by the state government (i.e.) Department of Agriculture, Horticulture, Animal Husbandry, Veterinary, Forestry, Fisheries etc.

Principles of Extension Education

1. Principle of interests and needs
2. Grass-root principle of organization
3. Principle of cultural difference
4. Principle of cultural change
5. Principle of co-operation & participation
6. Principle of applied science and democratic approach
7. Principle of learning by doing
8. Principle of trained specialists
9. Adoptability principle in the use of extension teaching methods

10. Principle of leadership
11. Principle of whole family
12. Principle of satisfaction

Principles of Extension Programme Planning

1. Extension programme should be based on an analysis of the past experiences, present situation and future needs.
2. Extension programme should have clear and significant objectives.
3. Extension programme should fix up priority on the basis of available resources and time.
4. Extension programme should clearly indicate the availability and utilization of resources.
5. Extension programme should have a general agreement at various levels.
6. Extension programme should involve people at the local level.
7. Extension programme should involve relevant institutions and organizations.
8. Extension programme should have definite plan of work.
9. Extension programme should provide for evaluation of results and reconsideration of the programme.
10. Extension programme should provide for equitable distribution of benefits amongst the members of the community.

Principles of Learning

1. Principles of self activity
2. Principles of association
3. Principles of transfer
4. Principles of disassociation
5. Principles of readiness
6. Principles of set or attitude
7. Principles of practice
8. Principles of motivation
9. Principles of timing
10. Principles of clarity of objectives
11. Principles of satisfyingness

Agricultural Technology Management Agencies (ATMA): It is a registered society responsible for technology dissemination at the district level. The Department of Agriculture and Cooperation, Govt. of India, intends to try a new model through establishment of Agriculture Technology Management Agencies (ATMA) in selected districts involving certain stages under the National Agricultural Technology Project (NATP). Launched by ICAR in 1999. It is a society of key stakeholders involved in agricultural activities for sustainable agriculture development in the district. The scheme is supported by the Central Government. The funding pattern is 90% by the central Government and 10% by the state government.

Agricultural Technology Management Agency (ATMA) /Governing Board

1. District Magistrate / Collector - Chairman
2. Chief Executive Officer / Chief Development Officer Vice - Chairman

Organizational Structure of ATMA

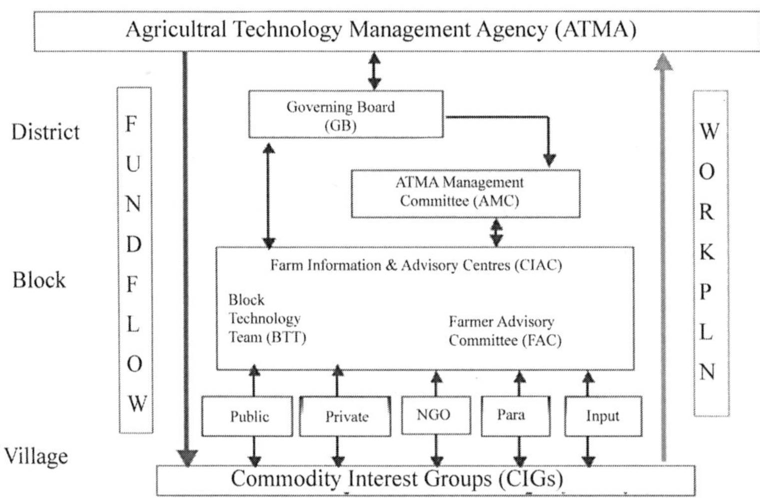

Agricultural Technology Information Center (ATIC, 1999): ATIC is a single window support system linking various unit of a research institution with intermediary and end users in decision making and problem solving exercise.

National Agricultural Innovation Project (NAIP): World bank assisted agriculture research programme, launched in India, July 2006.

National Agricultural Technology Project (NATP): This project was launched by the ICAR on June 30, 1998, with the support of the World Bank to strengthen and complement the existing resources and to argument the output National Agricultural Research System.

Krishi Vigyan Kendra: KVK is the nodal centre of farmers' training and technology transfer in agriculture and allied disciplines at the district level established by the Indian Council of Agricultural Research (ICAR). In KVK, subject matter specialists directly conduct training for farmers for their capacity building in various area related to Agriculture, Animal Husbandry, and Fishery etc.

On-Farm Testing (OFT): KVKs conduct OFT on proven agricultural technology at farmers' field so as to prove and fine-tune/modify the technology for its easy adoption by the farmers.

Front Line Demonstration (FLD): Front line demonstration is the demonstration of oilseeds, pulses, and other specific crops conducted by the KVKs in the farmers' field to prove the efficiency of the technologies to the farmers to help its adoption by them.

Step in Decision Making Process

1. Identification of the problem,
2. Finding alternatives
3. Finding solution,
4. Taking action to the plan,
5. Evaluation,
6. Responsibility.

Definition of learning

According to Rousseav (1953) "Learning is a process. It is changed in skills of human behaviour resulting from experience".

According to Colvin (1967) "Learning is the modifications of our readymade behaviour due to experience."

Learning Experience

A learning experience is defined as the mental or physical reaction one makes in a learning situation through seeing, hearing or performing activities during learning process/periods.

Learning theories

1. **Sensory stimulation theory:** This theory says that effective learning occurs when the senses are stimulated.

2. **Reinforcement Theory:** The theory was developed by the behaviorist school of psychology and it stresses that learner will repeat the desired behaviour if positive reinforcement follows the behaviour.
3. **Holistic Learning Theory:** Personality of an individual consists of many elements specially, the intellect, emotions, the body impulse, intuition, and imagination that all require activation if learning is to be more effective.
4. **Cognitive – Gestalt Approaches:** The emphasis here is on the importance of experience, meaning, problem solving and the development of insights.
5. **Facilitation Theory/Humanist Approach:** The root premise of this theory is that learning will occur by the educator acting as a facilitator through establishing an atmosphere in which learners feel comfortable to consider include ideas and are not threatened by external factors.
6. **Experiential Learning:** This theory asserts that without reflection people would simply continue to repeat their mistakes. The experiential learning cycle consist of concrete experience, conceptualization, and active experimentation.
7. **Action Learning:** Action learning is the approach that links the world of learning with the world of action through a reflective process within small cooperative learning groups known as 'action learning sets'.
8. **Adult Learning (Andragogy):** When adulthood arrives in the people they behave in adult ways and believe themselves to be adults. They feel they should be treated as adults.

Farmer Fields Schools: The Farmer Field School is a form of adult education, which evolved from the concept that farmers learn optimally from field observation and experimentation. Farmer Fields Schools came from the Indonesian expression Sekolah Lapangan. The first field school was established in 1989 in Central Java during a pilot season by 50 plant protection officers to test and develop field training methods.

Farm School: School established in the field of progressive/innovative farmers who are enhancing productivity and profitability in their farms through scientific and sustainable agriculture/animal husbandry practices. It is an alternative extension tool and facilitate farmer-to-farmer learning.

Entrepreneurship

Derived from French word Entreprendre - To undertake. French economist Richard Cantillon used the term entrepreneur to business.

According to Haggen "the function of seeing investment and production opportunity, organizing and enterprise to undertake a new production process, raising capital, hiring labour, arranging the supply of raw materials and selecting top managers for day-to-day operation of the enterprise."

Entrepreneur

According to McClelland entrepreneur is "some one who exercises some control over the means of production and produces more than what he can consume in order to sell it for profit. He executes the functions of coordination, organization and supervision. Entrepreneur is basically an innovator who introduces new combinations.

New encyclopedia Britannica defines "Entrepreneur is an individual who bears the risk of operating a business in face of uncertainty about the future conditions".

Intrapreneur

Intrapreneurs are usually employees within a company who are assigned a special idea or project, and are instructed to develop the project like an entrepreneur would. An intrapreneur acts independently, is proactive, and brings new ideas, which means that he brings innovations to the company. The intrapreneur's main job is to turn that special idea or project into a profitable venture for the company. Intrapreneurship is a concept of an employee thinking and acting as an entrepreneur, but within a company.

Market Intelligence: It is a process of giving you insights into what might happen in the near future market.

Domestic and Export Market Intelligence Cell (DEMIC): It is a cell which is disseminating timely, comprehensive, current and future price intelligence on agricultural commodities for better scientific decision making by farming community, traders, farmers and researchers.

Market Information Service (MIS): It is a means of increasing the efficiency of marketing systems and promoting improved price formation. Improved information enables farmers to plan their production more in line with market demand, schedule their harvests at the most profitable times, decide to which markets they should send their produce and negotiate the price with traders.

Contract Farming: Contract farming is a forward agreement between farmers and wholesalers/retailers for the production and supply of agricultural commodities and the agreement is generally made at predetermined prices.

APMC Act: Agricultural Produce Market Committee Act (APMC Act) is a State specific Act of law promulgated by the respective State Governments to regulate the marketing of agriculture produce.

Types of Entreprneurs

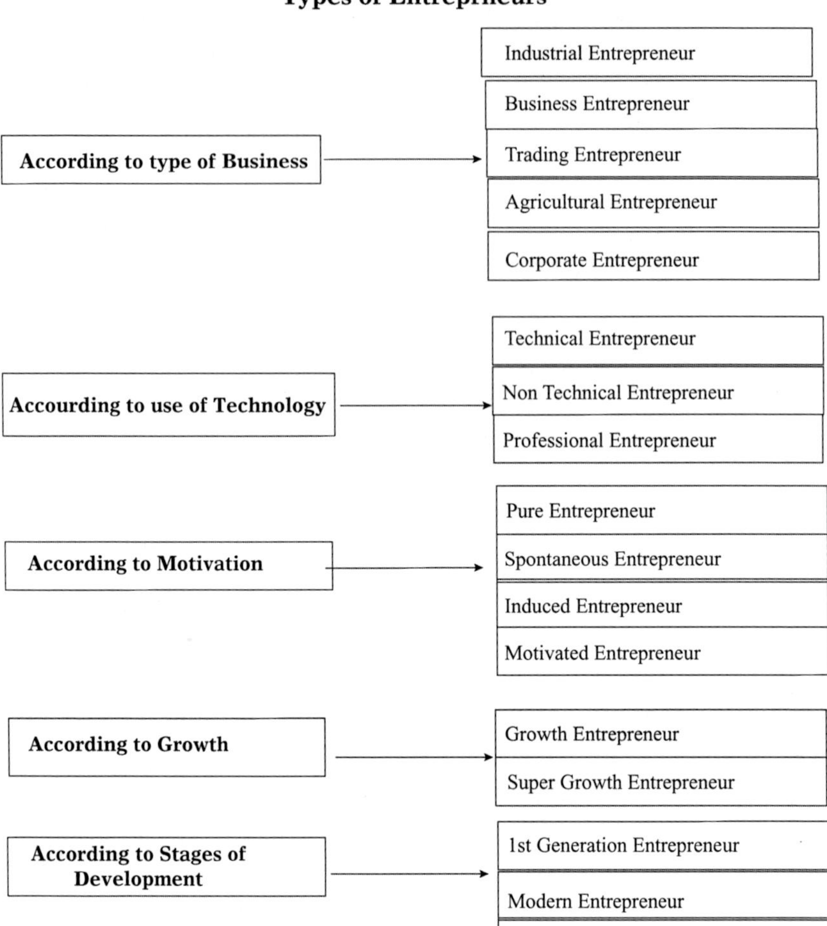

Agro processing: Defined as set of techno economic activities carried out for conservation and handling of agricultural produce and to make it usable as food, feed, fibre, fuel or industrial raw material. Hence, the agro-processing industry encompasses all operations from the stage of harvest till the material reaches the end users in the desired form.

Value addition: Value addition is the economically add value to a product and form characteristics more preferred in the market place. There are two main types of value addition, Innovation or coordination.

Supply Chain Management: Supply chain is a network of suppliers of raw materials, processing facilities and distribution channels. Supply Chain Management (SCM) is a set of practices & techniques to integrate the functioning of all the above 3 components for smooth & efficient flow of materials, information and money, to deliver the furnished product to customers.

Electronic Spot Exchange: This is a national level, institutionalized and transparent electronic exchange established by the National Commodity & Derivatives Exchange Ltd. (NCDEX) and is called the NCDEX Spot Exchange Ltd. (NCDEX Spot).

Eurep GAP: It is a private sector body that sets voluntary standards for the certification of agricultural products around the globe.

CODEX Alimentarius: It is a United Nations organization that establishes international food standards to protect consumer health and facilitate fair trade practices.

Marketing: It is a social and managerial process by which individuals and groups obtain what they want and need through creating, offering and exchanging products of value with others' - Philip Kotler

Market segmentation: Dividing a market into distinct subsets (segments) that behave in the same way or have similar needs."

Monopoly: Having only one supplier.

Oligopoly: In oligopolistic markets, independent suppliers (few in numbers and not necessarily acting in collusion) can effectively control the supply, and thus the price, thereby creating a seller's market. They offer largely similar products, differentiated mainly by heavy advertising and promotional expenditure, and can anticipate the effect of one another's marketing strategies. Examples include airline, automotive, banking, and petroleum markets.

Purchase: To buy a product or service.

Procurement: Procurement is the acquisition of goods, services or works from an outside external source. It is favorable that the goods, services or works are appropriate and that they are procured at the best possible cost to meet the needs of the purchaser in terms of quality and quantity, time, and location.

Acquisition: Complete process of obtaining goods and services—from preparation and processing of a requisition through to receipt and approval of the invoice for payment.

Raw Material: Basic substance in its natural, modified, or semi-processed state, used as an input to a production process for subsequent modification or transformation into a finished good.

Speculation: Deliberate assumption of above average short-term risk of financial loss, in expectation of above average gain from an anticipated change in prices. Organized speculation (conducted by commodity and stock exchanges) adds capital and liquidity to financial markets, and helps dampen wild fluctuations in prices in normal times.

Vendor: Manufacturer, Producer or Seller of good/ services.

Vertical Integration: Merger of firms at different stages of production and/or distribution in the same industry.

Backward integration: A firm acquires its input supplier.

Forward integration: A firms acquires in its output distribution chain.

Organic farming: A production system which avoids the use of synthetically manufactured fertilizers, pesticides, growth regulators and live stock feed additives. To the maximum extent feasible, organic farming systems rely on crop rotations, crop residues, animal manure, legumes, green manures, organic residual and aspects of biological pest control to maintain soil productivity and tilth, to supply plant nutrients and control insects, weeds and other pests.

Climate change: Defined as "Any long term substantial deviation from present climate because of variations in weather and climatic elements"

Indigenous Technical Knowledge (ITK): The local knowledge – knowledge that is unique to a given culture or society. ITK contrasts with the international knowledge system generated by universities, research institutions and private firms. It is the basis for local-level decision making in agriculture, animal husbandry, health care, food preparation, education, natural-resource management, and a host of other activities in rural communities.

Sustainable Rural Livelihoods: Livelihood options like farming, livestock rearing, forestry, etc. that ensure a satisfactory living for the rural families on a sustainable basis.

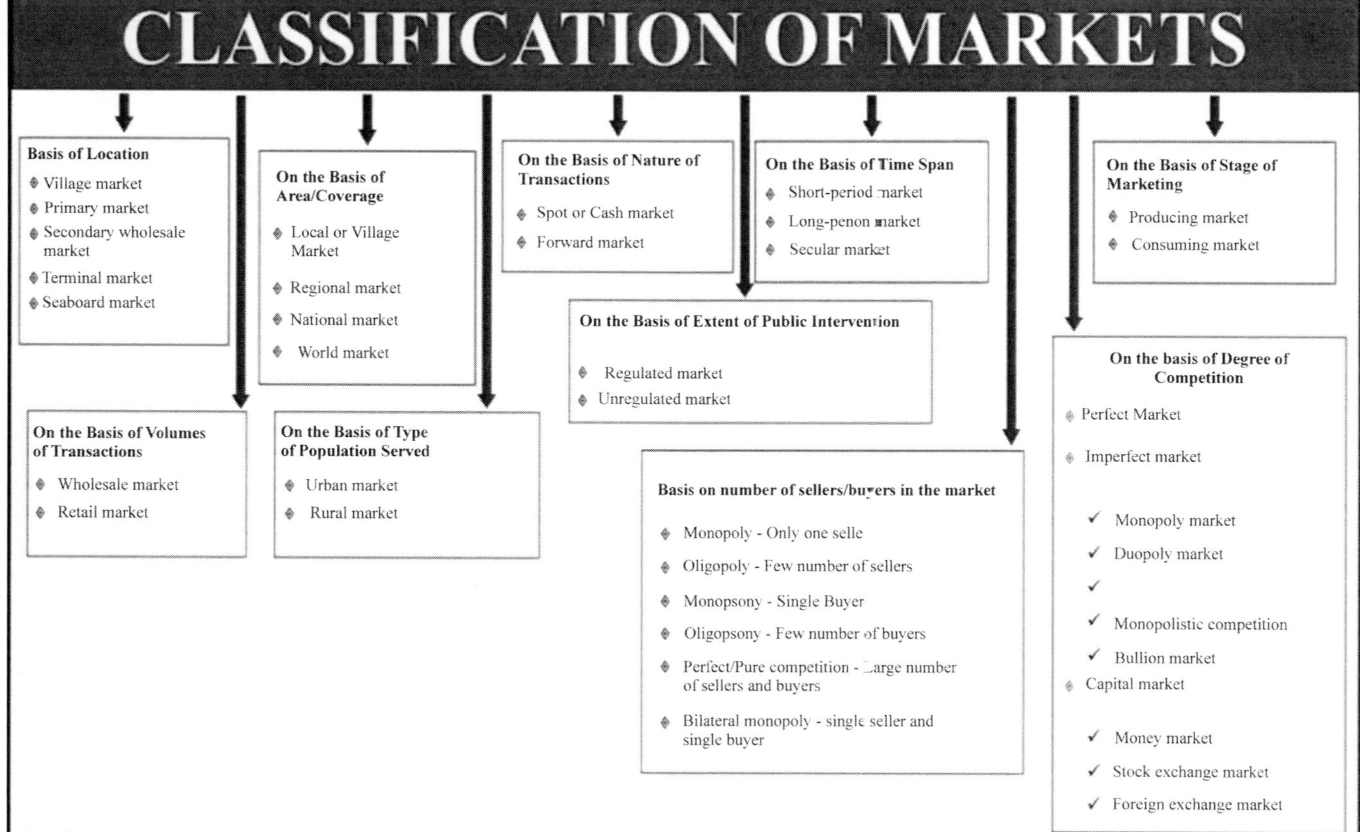
CLASSIFICATION OF MARKETS
Basis of Location
Village market
Primary market
Secondary wholesale market
Terminal market
Seaboard market
On the Basis of Volumes of Transactions
Wholesale market
Retail market
On the Basis of Area/Coverage
Local or Village Market
Regional market
National market
World market
On the Basis of Type of Population Served
Urban market
Rural market
On the Basis of Nature of Transactions
Spot or Cash market
Forward market
On the Basis of Time Span
Short-period market
Long-penon market
Secular market
On the Basis of Extent of Public Intervention
Regulated market
Unregulated market
Basis on number of sellers/buyers in the market
Monopoly - Only one selle
Oligopoly - Few number of sellers
Monopsony - Single Buyer
Oligopsony - Few number of buyers
Perfect/Pure competition - Large number of sellers and buyers
Bilateral monopoly - single seller and single buyer
On the Basis of Stage of Marketing
Producing market
Consuming market
On the basis of Degree of Competition
Perfect Market
Imperfect market
Monopoly market
Duopoly market
Monopolistic competition
Bullion market
Capital market
Money market
Stock exchange market
Foreign exchange market

Multiple Choice Questions

1. Entrepreneur word derived from:

   a) French c) Latin

   b) Greek d) None

2. The Entrepreneurship development programme having number of phases:

   a) One c) Four

   b) Six **d) Three**

3. Father of Extension education:

   a) K.N.Singh **c) James Stuart**

   b) Rogers d) None

4. An out-of-school system of education also known as:

   a) Communication b) Management

   c) Participation **d) Extension**

5. Extension means:

   a) Deciding for b) Expanding with

   c) Stretching out d) Thinking about

6. The term 'Extension' have root in:

   a) Latin b) Persian

   c) Greek d) Sanskrit

7. The Latin roots for the word 'extension' indicates:

   a) ex and termo b) extra and tension

   c) ex and luceo **d) ex and tensio**

8. The term "University Extension" was first used in the year:

   a) 1866 b) 1972

   c) 1901 d) 1921

9. The Binet test used for:

   a) Intelligence b) Attitude

   c) Knowledge d) None

10. The term "Extension Education" was first used in:

    a) USA b) India

    c) Netherlands **d) UK**

11. The word Extension first used by:

a) Vorhees b) Rogers

c) Aristotle d) Paul Leagans

12. Father of extension in India:

a) Adivi Reddy b) O P Dahama

c) K.N. Singh d) J B Chitambar

13. The word 'Vulgarisation' in French is equivalent to the word:

a) Social taboos b) Education

c) Extension d) Social change

14. Type of Extension aims at empowering or uplifting the poor:

a) Informative extension b) Formative extension

c) Emancipatory extension d) Persuasive extension

15. The word "Voorlichting" means:

a) Lighting the path b) Satisfaction

c) Improving skills d) Simplification

16. Father of demonstration in extension is:

a) James Stewatt b) Rober Chambers

c) G. De Thorde **d) Seamann A Knapp**

17. The book entitled "Extension Science" is written by:

a) N Roling b) P.Mathialagan

c) Van den Ban d) G L Ray

18. Extension as a science is come under:

a) Hard science b) Pure science

c) Soft science d) Physical science

19. Type of Extension that works on human resource development:

a) Emancipatory extension **b) Formative extension**

c) Informative extension d) Persuasive extension

20. Type of Extension used as a policy instrument for achieving societal objective:

a) Emancipatory extension b) Formative extension

c) Persuasive extension d) Informative extension

21. Farmers first model given by which scientist:
    a) Leagans **b) R Chambers**
    c) Van den Ban d) R M Rogers
22. Among them which of the fundamental objective of extension work:
    a) Farmer Development **b) Individual Development**
    c) Culture Development d) Society Development
23. Who was the author of book "Agricultural Extension":
    a) Van Den Ban b) R Chambers
    c) A Reddy d) R.M. Rogers
24. Extension education is an example of:
    a) Formal education b) Informal education
    c) Non-formal education d) All of the above
25. The basic unit of extension is:
    a) Family b) Society
    c) Individual d) Community
26. Agricultural extension in rural area deals with the:
    a) Farmers b) Change agents
    c) Farmers and change agents d) None of the above
27. With the help of extension worker, farmers attain desirable changes in his:
    a) Temperament **b) Behaviour**
    c) Health d) Wealth
28. Most useful leader in A.H. extension is:
    a) Democratic leader b) Bureaucratic leader
    c) Autocratic leader **d) Voluntary leader**
29. An extension worker is an example of:
    a) Professional leader b) Politician
    c) Local leader d) Voluntary leader
30. Which type of motivation important for extension:
    a) Extrinsic motivation **b) Intrinsic motivation**
    c) Situational motivation d) Classical motivation

31. Indian Research Journal of Extension Education is published from:

a) SEEA, Agra b) GBPUAT, Pantnagar

c) TNAU, Coimbatore d) Directorate of Extension

32. The Science of adult learning is:

a) Androgogy b) Pedagogy

c) Pedology d) Edaphology

33. The word androgogy have originated from:

a) Latin **b) Greek**

c) Sanskrit d) German

34. In andrological method emphasis given on:

a) Content and Process b) Content

c) Process d) None of the above

35. "Here and Now" condition is the characteristic feature of:

a) Adult learning b) Child learning

c) Edagogy d) None

36. The term "Andragogy" was given by:

a) P Freire b) Edger dale

c) R Chambers **d) Alexander Kapp**

37. The central problem of adult learning is:

a) Age **b) Motivation**

c) Sex difference d) Literacy

38. The term pedagogy has its origin from:

a) French **b) Greek**

c) Latin d) Sanskrit

39. "Learning the rope" indicates:

a) Socialization b) Training

b) Both a & b d) None of the above

40. The 'extension education process' concept was given by:

a) Paul Leagans b) Curt Lewin

c) Ensminger d) Wilson and Galup

41. Among which of the following last stage of extension education process is:

    a) Evaluation **b) Reconsideration**

    c) Adoption d) Teaching

42. Extension Education is essentially produced desirable change in:

    a) Knowledge of farmers b) Attitude of farmers

    c) Practice of farmers **d) Behavior of farmers**

43. World literacy day was celebrated on:

    a) September 15 **b) September 8**

    c) June 9 d) February 21

44. National Women Literacy Day is celebrated on:

    a) September 8 **b) October 2**

    c) November 2 d) August 15

45. The steps in extension teaching given by:

    a) Wilson & Gallup b) Ensminger

    c) Curt Lewin d) Paul Leagans

46. Extension teaching steps are:

    a) AIDCAS b) AIETA

    c) KPDC d) AIDACS

47. During extension teaching the teacher should know the:

    a) Subject matter b) Teaching aids

    c) Place of teaching **d) Attitude of the learner**

48. Learning process mainly controlled by:

    a) Teacher **b) Learner**

    c) Subject matter d) Teaching materials

49. Bond theory of learning was given by:

    a) O.P. Dhama b) Egder Dale

    c) E L Thorndike d) None of these

50. The normal shape of learning curve is:

    a) S-shape b) U-shape

    c) V-shape d) O-shape

51. Father of learning is:

a) Egder Dale | b) Aristotle
c) Thorndike | d) Knapp

52. The adult education phenomenon first given by:

a) Robert Davis | b) Paul Leagans
c) Thomas Pale | d) K N Singh

53. Cone of experience classifies the instructional methods according to:

a) Experimental correctness | b) Difficulty in understanding
c) Easy to use and application | d) Level of abstraction in thinking

54. Cone of experience was given by scientist:

a) Berlo C K | b) Rogers
c) Egder Dale | d) S C Parmer

55. In which country project 'Schools on the air' is working:

a) Germany | b) Bangladesh
c) Philipines | d) Srilanka

56. In which country Exteesion workers are called as "agricultural representatives":

a) Canada | b) UK
c) USA | d) Netherlands

57. "Rural Agent" known as extension worker in:

a) Mexico | **b) Peru**
c) USA | d) Netherlands

58. Where the first agricultural university was established:

a) Pantnagar | b) Coimbatore
c) Darward | d) Cuttack

59. Extension teaching at under graduate level was started by:

a) Madras University | b) Meerut University
c) Calcutta University | d) Mumbai University

60. Indian Society of Extension Education was established on:

a) 11th September 1962 | b) 18th November 1963
c) 22nd June 1964 | d) 17th January 1965

61. The first extension education institute (EEI) in India is:

a) Nilokari b) Anand

c) Hyderabad d) Jorhat

62. ASRB was established in the year:

a) 1971 **b) 1973**

c) 1975 d) 1979

63. The first line system of extension in India is controlled by:

a) ICAR institutions/SAUs b) Ministry of Agriculture

c) Non Government Organizations d) Private Extension Agencies

64. In which year division of Extension in ICAR was established:

a) 1924 b) 1954

c) 1971 d) 1977

65. Result demonstration was conducted in:

a) Research Station **b) Farmers field**

c) District Farms d) None of the above

66. The principle of result demonstration was:

a) Learning is believing b) Learning by teaching

c) Seeing is believing d) Learning by doing

67. Methods of motivating the farmers by showing that what has actually been achieved by applying the practice under field conditions.

a) Campaign b) Result demonstration

c) Field day d) Farm bulletin

68. Obliteration of group boundaries happened in:

a) Method demonstration b) Farm and Home visit

c) Office calls **d) Campaign**

69. Puppet show is an example of:

a) Mass media b) Group media

c) Individual media d) Tribal media

70. The book entitled "Traditional Media in India" is written by:

a) Aristotle b) K.N.Singh

c) D. Nair **d) S. C. Parmer**

71. A letter written by the extension worker to farmers, related with extension work of an area known as:

a) News letter | **b) Circular letter**
c) Personal letter | d) None of the letter

72. A systematic display of models, specimens charts, etc. in a sequence around a theme.

a. Campaign | b) Farmers fair
c) Demonstration | **d) Exhibition**

73. Most frequently used mass media by rural community:

a) Television | **b) Radio**
c) Films | d) Print Media

74. In a field of knowledge, body of general principles is known as:

a) Objectives | **b) Philosophy**
c) Mandate | d) Principle

75. The term philosophy has roots in:

a) Greek | b) French
c) Latin | d) Sanskrit

76. The basic philosophy of extension work is to teach people:

a) How to think | b) What to think
c) When to do | d) Where to do

77. Who given the principles of extension service:

a) M Horton | b) P Leaganes
c) Ensminger | d) O P Dahama

78. Establishment of Panchayat Raj Bodies at various level fulfils the extension principle of:

a) Cooperation | b) Leadership
c) Participation | **d) Grass root**

79. Regular advertisement regarding public health and vaccination at the onset of an epidemic outbreak is a example of:

a) Principle of set | **b) Principle of time**
c) Principle of association | d) Principle of transfer

80. In India, extension approach in follows the norms of:

a) Prescription | b) Participation
c) Cooperation | d) Mobilization

81. The preferred approach in extension work is:
    a) Autocratic **b) Democratic**
    c) Lessiz faire d) None of the above
82. Among which of the privatization approach, public pays and private delivers:
    a) Contracting b) Farmers association
    c) Self help **d) Vouchers**
83. The magazine 'Kheti' started in the year:
    a) 1929 b) 1930
    c) 1938 d) 1948
84. In learning situations who got the central position:
    a) Learner b) Teacher
    c) Both a & b d) None
85. Example of bottom up approaches is:
    a) Farming system approach
    b) Commodity specialized approach
    c) Training and Visit approach
    d) Coast sharing approach
86. The difference between what is and what ought to be known as:
    a) Need b) Aim
    c) Goal d) Objective
87. The condition which people after study decided to change is:
    a) Proble b) Programme
    c) Project d) Solution
88. A blue print of action is known as:
    a) Plan b) Programme
    c) Objective d) Project
89. The distance in any given direction to go is known as:
    a) Aim **b) Goal**
    c) Objective d) Principle
90. For change an unsatisfactory situation, course of proposed action is known as:
    a) Solution b) Situation
    b) Objective d) Aim

91. Procedure of attaining an objective is called:
   a) Programme **b) Method**
   c) Plan d) Media

92. The expression of ends towards which our efforts are directed known as:
   a) Aim b) Goal
   c) Objective d) Target

93. In an extension work, objective operates at field level is known as:
   a) Fundamental objective **b) Working objective**
   c) General objective d) None of the above

94. Which among the following is not a level of objective:
   a) Basic objective b) Fundamental objective
   c) Working objective d) General objective

95. In a project, who, what, when and where are the part of:
   a) Plan of the work b) Calendar of the work
   c) Programme evolution d) Decision making

96. When plan of work arranged chronologically is called:
   a) Calendar of work b) Calendar of operation
   c) Objective d) Porgramme

97. A sound programme must have character like:
   a) Permanence with rigidity
   b) Permanence with flexibility
   c) Temporarily
   d) Temporarily with flexibility

98. Extension programme planning is a:
   a) Continuous process b) Discontinuous process
   c) Effective process d) Uniform process

99. Extension work grew out of:
   a) Problem of an area **b) Situation of an area**
   c) Condition of an area d) Participation of people

100. Felt and unfelt type of need is classified on the basis of:
   a) Physical point of view **b) Psychological point of view**
   c) Social point of view d) All of the above

101. The first step in extension programme planning is:

a) Determination of objectives | **b) Analysis of existing situation**
c) Determination of target | d) Decision of extension teaching method

102. According to Kelsey and Hearne the number of steps in programme planning is:

a) Four | **b) Seven**
c) Three | d) Six

103. The participation of people in extension programme is:

a) Voluntary | b) Compulsory
c) Involuntary | d) None of the above

104. The extension service in USA is called:

a) National extension service | b) U S Extension work
c) Cooperative extension service | d) Rural extension service

105. Chairman of the steering Committee of 'Global Forum for Rural Advisory Services' (GFRASS)?

a) Rasheed Sulaiman | b) Sulekha Dev Burman
c) Andy Hall | d) Van den Ban

106. CEO of Global Youth Innovation Network (GYIK)?

a) Pape Samb | b) Hugo Douglas
c) Nia arnolin | d) None of these

107. Which of the following aids are called 'Spark Plugs'?

a) Maps | b) Charts
c) Graphs | **d) All of theses**

108. Under which of the following taxonomy hierarchy of six levels of cognition – knowledge, comprehension, application, analysis, and evaluation, appears?

a) Linnaeus Taxonomy | **b) Bloom's Taxonomy**
c) Karanz Taxonomy | d) All of theses

109. Long lasting results can be seen if extension focuses more on ____________

a) Input supply | **b) Output**
c) Credit supply | d) All of theses

110. Theory of Cognitive Dissonance was given by

a) Richard Lazarus
b) Leon Festinger
c) Jean Piaget's
d) None of theses

111. The market – led extension farmers known as

a) Innovation
b) Entrepreneurs
c) High producers
d) Agripreneurs

112. ITK is

a) Holistic
b) Qualitative and practical
c) Experience
d) All of the above

113. How to do vaccination ____________?

a) National demonstration
b) Result demonstration
c) Methods demonstration
d) None of theses

2

Women Empowerment (Theory)

The Constitution of India related to women

- Guarantees to all Indian women equality: Article 14
- No discrimination by the State: Article 15(1)
- Equality of opportunity: Article 16
- Equal pay for equal work: Article 39(d)
- Securing just and humane conditions of work and for maternity relief: Article 42
- The All India Women's Education Conference was held in Pune in 1927.
- The National Policy For the Empowerment of Women came was passed in 2001
- In 2010 March 9, Rajya Sabha passed the Women's Reservation Bill requiring that 33% of seats in India's Parliament and state legislative bodies be reserved for women.
- Women perform 2/3 of the world's work
- Women earn 1/10 of the world's income
- Women comprise 2/3 of the world's illiterate
- Women own less than 1/100 of the world's property

International Women Development Agency: Melbourne (Australia)

The National Commission for Women: Set up as a statutory body in January 1992 under the National Commission for Women Act, 1990 (Act No. 20 of 1990) of Govt. of India. The Year 2001 was announced by the Indian government as Women's Empowerment Year.

Women Empowerment

Women's empowerment is very essential for the development of society. Empowerment means individuals acquiring the power to think and act freely, exercises choice and fulfill their potential as full and equal members of society.

According to Friedmann's 1992 empowerment involves local self-reliance, direct participatory democracy and experiential social learning. All three kinds of empowerment are seen as relevant to women's struggles and form an interconnecting triad. When this triad, centered on an individual woman and household, is linked up with others, the result is a social network of empowering relations that, because it is mutually reinforcing, has extraordinary potential for social change.

According to Rappaport 1987 defined empowerment as "a process by which people, organizations, and communities gain mastery over issues of concern to them". In the simplest of words it is basically the creation of an environment where women can make independent decisions on their personal development as well as shine as equals in society.

Indicators of Women Empowerment

Quantitative indicators: It measure quantifiable results. Eg: school enrollment rates or percentage of women in parliament. This data is often drawn from censuses, administrative records, and other large-scale surveys.

Qualitative indicators: Data show differences that cannot be described numerically, such as differences in quality levels of policies. Qualitative indicators permit more in-depth examination of social processes, social relations, power dynamics, and gender equity. These indicators capture perceptions, opinions, and experiences. They can be collected through focus groups, social mapping, interviews, and surveys.

Both quantitative and qualitative indicators can measure gender differences with two general approaches. One is to measure distinctions between males and females only (e.g., the number of girls enrolled in school compared to boys). The other is to assess the impact of gender equality on overall well-being of population, as gender inequality affects not only individual women, but society as a whole.

Gender Empowerment Measure: The GEM was measure the "whether women and men are able to actively participate in economic and political life and take part in decision-making" (UNDP, 1995). The GEM is determined using three basic indicators:

1. Proportion of seats held by women in national parliaments,
2. Percentage of women in economic decision making positions (incl. administrative, managerial, professional and technical occupations)
3. Female share of income (earned incomes of males vs. females).

The Gender Parity Index (GPI): It is a socioeconomic index usually designed to measure the relative access to education of males and females. It is used by international organizations, particularly in measuring the progress of developing countries.

Gender: Gender is generally confused with sex.

Sex: Biological term which refers to the functional differences between males and females and their reproductive potential. It can further be defined as- genes in chromosomes.

Gender: A neutral term. It is a socially constructed roles and responsibilities assigned to men and women in a given society.

Gender Bias: The tendency to make decisions or take actions based on gender.

Gender role: It can be defined as roles which are classified by sex. It can be identified as behaviour or tasks done with some expectation or self-consciousness about being female or male. Gender is a dynamic concept – gender roles for women and men differ from one culture to another and from one social group to another within the same culture. Race, Class, economic circumstances, age – all of this influence what is considered appropriate for women and men. Furthermore, as culture is dynamic and socio-economic conditions change over time, so gender roles change with them.

Gender awareness: Gender awareness is an understanding that there are socially determined differences between women & men based on learned behaviour, which affect their ability to access and control resources. This awareness needs to be applied through gender analysis into projects, programmes and policies.

Gender-Equality: The right of equality is a fundamental one. Gender equality also known as sex equality, sexual equality or equality of the genders, refers to the view that men and women should receive equal treatment, and should not be discriminated against based on gender, unless there is a sound biological reason for different treatment. Gender equality means that there is no discrimination on grounds of a person's sex in the allocation of resources or benefits, or in the access to services.

A **"gender-equal society"** is a "society in which both men and women, as equal members, have the opportunity to participate in all kinds of social activities at will, equally enjoy political, economical and cultural benefits, and share responsibilities".

Gender inequality: Unequal treatment or perceptions of individuals based on their gender. It arises from differences in socially constructed gender roles as well as biologically through chromosomes, brain structure, and hormonal differences. Gender systems are often dichotomous and hierarchical.

Gender equity: Gender equity entails the provision of fairness and justice in the distribution of benefits and responsibilities between women and men. The concept recognizes that women and men have different needs and power and that these differences should be identified and addressed in a manner that rectifies the imbalances between the sexes.

Gender discrimination: Prejudice or discrimination based on a person's sex or gender. Sexist attitudes may include the belief that a person of one sex is intrinsically superior to a person of the other. OR Prejudicial treatment of an individual based on a gender stereotype (often referred to as sexism or sexual discrimination).

Gender sensitization: Modification of behaviour by raising awareness of gender equality concerns. Gender sensitization theories claim that modification of the behavior of teachers and parents (etc.) towards children can have a causal effect on gender equality.

Gender differentiation: Results in gender inequality, with one gender being regarded as inferior to the other with regards to certain activities.

Gender mainstreaming: Mainstreaming essentially offers a pluralistic approach that values the diversity among both women and men. The concept of gender mainstreaming was first proposed at the 1985 Third World Conference on Women in Nairobi, Kenya. The idea has been developed in the United Nations development community. Gender mainstreaming is a strategy for integrating gender concerns in the analysis, formulation and monitoring of policies, programmes and projects. It is therefore a means to an end, not an end in itself; a process, not a goal. The purpose of gender mainstreaming is to promote gender equality and the empowerment of women in population and development activities OR Gender mainstreaming is the process of ensuring that women and men have equal access and control over resources, development benefits and decision-making, at all stages of the development process and projects, programmes and policy.

Gender-blind: Gender blindness is the failure to recognize that gender is an essential determinant of social outcomes impacting on projects and policies.

Gender Sensitive: Being aware of the differences between women's and men's needs, roles, responsibilities, and constraints.

Gender-sensitivity: Gender sensitivity encompasses the ability to acknowledge and highlight existing gender differences, issues and inequalities and incorporates these into strategies and actions.

Gender issues: Specific consequences of the inequality of women and men.

Gender Relations: The ways in which a culture or society defines rights, responsibilities, and identities of men and women in relation to one another.

Gender Division of Labour: The roles, responsibilities, and activities assigned to women and men based on gender.

Gender Analysis: Gender Analysis is the process of analyzing information in order to ensure development benefits and resources are effectively and equitably targeted to both women and men, and to successfully anticipate and avoid any negative impacts development may have on women or on gender relations.

Gender Budgeting: Gender Budgeting is a dissection of the Government budget to establish its gender-differential impacts and to translate gender commitments into budgetary commitments. Thus Gender Budgeting looks at the Government budget from a gender perspective to assess how it addresses the needs of women in the areas like health, education, employment, etc. Gender Budgeting does seeks affirmative action to address specific needs of women.

Difference between Gender equity and Gender equality

Gender equity	Gender equality
Gender equity is the process of being fair to women and men.	Gender equality requires equal enjoyment by women and men of socially-valued goods, opportunities, resources and rewards.
Equity leads to equality.	In other words, gender equality refers to equal access to social goods, services and resources and equal opportunities in all spheres of life for both men and women
Equity is the quality of being fair or impartial.	Equality is the state of being equal.
Equity is essential to achieve true equality.	Achieving gender equality requires women's empowerment to ensure that decision-making at private and public levels, and access to resources are no longer weighted in men's favour, so that both women and men can fully participate as equal partners in productive and reproductive life.

Programmes for Women Empowerment

Millennium Development Goal

The United Nations Development Programme constituted eight Millennium Development Goals (MDG) for ensuring equity and peace across the world. The third MDG is directly related to the empowerment of women in India. The MDGs are agreed-upon goals to reduce certain indicators of disparity across the world by the year 2015.

The third MDG is centred towards promoting gender equality and empowering women and state that: "Eliminate gender disparity in primary and secondary education, preferably by 2005, and in all levels of education by no later than 2015". However, the country has hastened progress and the Gender Parity Index (GPI) for Gross Enrolment Ratios (GER) in primary and secondary education has risen.

Swayamsidha Programme

Swayamsidha scheme was launched in the year 2001 dedicated to Women's Empowerment. It is a Self Help Groups based programme with emphasis on convergence activities.

Swayamsidha programme – an integrated scheme for the empowerment of women at a total cost of Rs. 116.30 Crores. Core to this programme will be the establishment of women's self-help groups which will empower women to have increased access to all kinds of resources that they are denied, in addition to increasing their awareness and skills. This programme will benefit about 9,30,000 women with the setting up of 53,000 self-help groups, 26,500 village societies and 650 block societies.

Mother and Child Tracking System (MCTS)

Ministry of Health and Family Welfare launched the Mother and Child Tracking System (MCTS) in December 2009.

The Indira Gandhi Matritva Sahyog Yojana Conditional Maternity Benefit Plan (CMB)

Indira Gandhi Matritva Sahyog Yojana (IGMSY), Conditional Maternity Benefit (CMB) is a centrally sponsored scheme for pregnant and lactating (P & L) women to improve their health & nutrition status by providing cash incentives to pregnant and nursing mothers. The scheme was introduced in October, 2010. The scheme is 100% centrally sponsored and funds are released from Government of India to States/UTs as prescribed under scheme guidelines.

Rajiv Gandhi scheme for empowerment of adolescent girls – Sabla (RGSEAG)

The Rajiv Gandhi Scheme for Empowerment of Adolescent Girls - Sabla (RGSEAG) was initiated on April 1, 2011 by the Ministry of Women and Child Development. The adolescent girls aged 10–19 years old constitute almost 47% of the total population of adolescents in India. The school or educational drop-out rate among the girls is quite high. Adolescent girls have a drop-out rate of 21% and adolescent boys have a drop-out rate of 8%. Dropout rates among girls are high largely due to the distance from their homes to the schools, predominance of male teachers, poor sanitation facilities at school, the incidence of early marriage, and the early assumption of domestic responsibilities in the home.

Indira Mahila Yojana

Another women development programme intiated by the Government, which was launched in August 1995 in more than 200 blocks of the country. The main objective of this programme is to give a forward thrust to the women education, awareness income-generation capacities and the empowerment of women. The platforms for the forward thrust are to be the self-help groups at the gross-root level.

The Rashtriya Mahila Kosh

Rashtriya Mahila Kosh (The National Credit Fund for Women), referred to as RMK, was set up in 1993 to meet the credit needs of poor women in the informal sector for income generation and household activities. Established by the Department of Women and Child Development, under the Ministry of Human Resource Development, for the poor women from both rural and urban areas whose family income does not exceed Rs. 11,000/- per annum in rural areas and Rs. 11,800/- per annum in urban areas. The interest charged is 12% per annum to the ultimate borrower women and 8% to the NGOs and the corporations. The RMK was originally set up under the Societies Act in 1860, and is managed by a governing board comprising 16 members.

The Velugu Programme

The Government of Andhra Pradesh has been implementing since June 2000 a special project called "Velugu" to address poverty through empowerment of rural poor women, especially the poorest of the poor. The project is being implemented by the Society for Elimination of Rural Poverty and the Panchayati Raj, Government of Andhra Pradesh.

Sukanya Samridhi Yojna

Sukanya Samriddhi yojna is a small deposit scheme for girl child, as part of 'Beti Bachao Beti Padhao' campaign, which would fetch yearly interest rate of 9.1 per cent. The scheme was operate with the name Sukanya Samriddhi Account, from December 2014.

Policies and Schemes for Women Entrepreneurs in India

In India, the Micro, Small & Medium Enterprises development organisations, various State Small Industries Development Corporations, the Nationalised banks and even NGOs are conducting various programmes including Entrepreneurship Development Programmes (EDPs) to cater to the needs of potential women entrepreneurs, who may not have adequate educational background and skills. Small Industries Development Bank of India (SIDBI) has also been implementing special schemes for women entrepreneurs.

- By Ministry of MSME
  - Trade related entrepreneurship assistance and development (TREAD) scheme for women
  - Mahila Coir Yojana
- By Ministry of Women and Child Development
  - Support to Training and Employment Programme for Women (STEP)
  - Swayam Siddha
- By Kerala State Women's Development Corporation
  - Self employment loan programmes
  - Educational loan schemes
  - Single women benefit schemes
  - Job oriented training programmes
  - Marketing support for women entrepreneurs
  - Autorickshaw / school van's driver scheme

Women Entrepreneur Associations

1. Federation of Indian Women Entrepreneurs (FIWE)
2. Consortium of Women Entrepreneurs (CWEI)
3. Association of Lady Entrepreneurs of Andhra Pradesh
4. Association of Women Entrepreneurs of Karnataka (AWAKE)
5. Self-Employed Women's Association (SEWA)
6. Women Entrepreneurs Promotion Association (WEPA)
7. The Marketing Organisation of Women Enterprises (MOOWES)
8. Bihar Mahila Udyog SanghBihar Mahila Udyog Sangh
9. Mahakaushal Association of Woman Entrepreneurs (MAWE)
10. SAARC Chamber Women Entrepreneurship Council

11. Women Entrepreneurs Association of Tamil Nadu (WEAT)
12. TiE Stree Shakti (TSS)
13. Women Empowerment Corporation

CDR Environments of Robert Chambers: According to Robert's vision of poverty is not the poverty of simple GDP figures or poverty lines. He has focused always on the individuals who underlie the statistics: on their isolation, their powerlessness, their vulnerability to shocks like illness, their coping strategies. He has emphasized that the poor are found mostly in environments which are complex, diverse and risk prone, the so called CDR environments. He has focused not simply on increasing income, but rather on the empowerment of the poor, on a reduction in their vulnerability, on helping them to exploit the diversity of their environments.

Multiple Choice Questions

1. Guarantees to all Indian women equality mentioned in which article:
   a) Article 14
   b) Article 15(1)
   c) Article 16
   d) Article 42
2. No discrimination by the State say in which article:
   a) Article 14
   b) Article 15(1)
   c) Article 16
   d) Article 42
3. Equality of opportunity for both men and women mentioned in which article:
   a) Article 14
   b) Article 15(1)
   c) Article 16
   d) Article 42
4. Equal pay for equal work mentioned in which article:
   a) Article 14
   b) Article 39(d)
   c) Article 16
   d) Article 42
5. Securing just and humane conditions of work and for maternity relief mentioned in which article:
   a) Article 14
   b) Article 15(1)
   c) Article 16
   d) Article 42
6. The National Policy For the Empowerment of Women was passed in the years:
   a) 2002
   b) 2001
   c) 2003
   d) 2005

7. Women's Reservation Bill passed in the years:

   a) 9th March 2010 c) 9th March 2008

   b) 8th April 2011 d) 12th June 2010

8. India's parliament and state legislative bodies reserve the percentage of seats for women:

   a) 21% **c) 33%**

   b) 50% d) 29%

9. Women perform ---- of the word's work

   a) 2/3rd c) 1/10th

   b) 3/4th d) 1/3rd

10. Women earn ---- of the word's income

    a) 2/3rd **c) 1/10th**

    b) 3/4th d) 1/3rd

11. National Commission for Women set up in the years:

    a) 1992 c) 1991

    b) 1990 d) 1997

12. National Commission for Women is a:

    a) Autonomous body **c) Statutory body**

    b) Private d) None

13. Which Millennium development goal directly related to empowerment of women in India:

    a) Ist c) IInd

    b) IIIrd d) Vth

14. The term 'Gender' refers to

    a) Culturally based expectation c) Focused only on women

    b) Rules followed by community d) None

15. The number of member in SHGs

    a) 10-20 c) 5-10

    b) 5-20 d) None

16. NRCWA situated at:

    a) New Delh c) Mumbai

    b) Bhubaneswar d) Ludhiana

17. No discrimination on grounds of a person's sex in the allocation of resources or benefits known as:

    a) Gender Equality
    b) Gender Role
    c) Gender Equity
    d) Gender awareness

18. Role which are classified by sex:

    a) Gender Differentiation
    b) Gender Role
    c) Gender Discrimination
    d) All

19. Provision of fairness and justice in the distribution of benefits and responsibilities between women and men commonly known as:

    a) Gender Equity
    b) Gender bias
    c) Gender Equality
    d) None

20. Prejudice based on a person's sex or gender:

    a) Gender Discrimination
    b) Both a & c
    c) Gender Inequality
    d) None

21. Gender mainstreaming is a:

    a) Means to an end
    b) Both a & c
    c) Process
    d) None

22. Concept of gender mainstreaming first proposed in:

    a) IIIrd World Conference on Women in Nairobi, 1985
    b) Ist World Conference on Women in Mexico city, 1975
    c) IInd World Conference on Women in Copenhagen, 1980
    d) IVth World Conference on Women in Beijing, 1995

23. International Women's Year (IWY) was the name given to Year:

    a) 1975
    b) 1990
    c) 1986
    d) 1995

24. Women's Empowerment Year is:

    a) 2004
    b) 2006
    c) 2001
    d) 2011

25. Mother and Child Tracking System launched in the year:

    a) 2011
    b) 2001
    c) 2009
    d) 2010

26. GEM indicates:
    a) Genetic Element Measure
    b) Gender Empowerment Measure
    c) Gender Empowerment Measurement
    d) None
27. Socioeconomic index designed to measure the relative access to education of males and females.
    a) Gender Empowerment Measure
    c) Gender Parity Index
    b) None
    d) Gender equality index
28. The roles, responsibilities, and activities assigned to women and men based on gender.
    a) Gender Discrimination
    c) Gender Role
    b) Gender division of labour
    d) All
29. Swayamsidha scheme was launched in the year:
    a) 2002
    c) 2001
    b) 2003
    d) 2004
30. Swayamsidha scheme is a:
    a) Self Help Group based programme
    b) Community Group based programme
    c) Formal Group based programme
    d) None
31. Indira Gandhi Matritva Sahyog Yojana (IGMSY) is a:
    a) Centrally sponsored scheme
    b) State Govt. sponsored scheme
    c) Both
    d) None
32. Indira Gandhi Matritva Sahyog Yojana (IGMSY) launched in the year:
    a) 2009
    c) 2011
    b) 2010
    d) 2008
33. Contribution of Govt. of India in IGMSY:
    a) 50%
    c) 100%
    b) 33%
    d) None
34. The Rajiv Gandhi Scheme for Empowerment of Adolescent Girls - Sabla (RGSEAG) was initiated on
    a) April 1, 2011
    c) April 1, 2009
    b) April 1, 2012
    d) April 1, 2000

35. Drop-out rate of Indian adolescent girls:

a) 25% c) 20%

b) 21% d) 33%

36. Indira Mahila Yojana intiated by the Government, in which year:

a) August 1995 c) June 1997

b) July 2000 d) None

37. Rashtriya Mahila Kosh set up in the year:

a) 1994 **c) 1993**

b) 1995 d) 1997

38. Credit needs of poor women in the informal sector for income generation and household activities meet by:

a) Indira Mahila Yojana **c) Rashtriya Mahila Kosh**

b) Swayamsidha d) None

39. Rashtriya Mahila Kosh governing board comprising:

a) 16 members c) 20 members

b) 10 members d) 30 members

40. Velugu Programme launched in the year:

a) 2000 c) 2001

b) 2002 d) 2003

41. Women workforce participation according to National Sample Survey Organization (2011-12) was

a) 45% **c) 22.5%**

b) 21% d) 31%

42. Highest women workforce participation in which Country:

a) America c) England

b) Norway d) China

43. The concept of Grameen Bank in Bangladesh is popularized by:

a) Mohd. Younis b) Aftab Hussain

c) Aga Khan d) Zabir Pasha

44. From which country concept of five year plan was adopted:

a) China **b) Russia**

c) America d) Britain

45. The article under WTO that deals with agricultural sector trade reforms:
    a) Article 20 b) Article 15
    c) Article 10 d) Article 5
46. In which five year plan removal of poverty was the main objective:
    a) Fourth **b) Fifth**
    c) Sixth d) Seventh
47. The term "Hindu rate of growth" was given by:
    a) K N Singh b) K N Raj
    c) Raj Krishna d) Mahalanobis
48. In which five year plan, India achieved self reliance in food production:
    a) Fourth plan b) Third plan
    c) Second plan d) Seventh plan
49. Micro level Planning is come under:
    a) Imperative planning b) Five year planning
    c) Regional planning d) Target group planning
50. Five year plan is an example of:
    a) Perspective plan **b) Medium term plan**
    c) Adhoc plan d) Contingent plan
51. A rolling plan is run for:
    a) One year b) Two years
    c) Three years **d) Year to year basis**
52. The current five year plan is:
    a) Ninth plan b) Eleventh plan
    c) Tenth plan **d) Twelfth plan**
53. Among them which programme provide the housing needs of the poor:
    a) IAY b) JRY
    c) MWS d) NREP
54. The construction assistance under the IAY is:
    a) Rs. 25,000/- per unit **b) Rs. 35,000/- per unit**
    c) Rs. 45,000/- per unit d) Rs. 55,000/- per unit
55. Bharat Nirman project covers all the areas except:
    a) Rural connecativity b) Rural housing
    c) Rural water supply **d) Primary education**

56. The targets aimed under Bharat Nirman is to be achieved by the year:

a) 2009 b) 2011

c) 2008 d) 2015

57. As per 2011 census of India, density of population is:

a) 267 per km^2 **b) 382 per km^2**

c) 356 per km^2 d) 388 per km^2

58. The year in which 'National Drinking Water Mission' was launched:

a) 1975 **b) 1986**

c) 1991 d) 1996

59. The year in which 'Million well scheme' was launched:

a) 1988-89 b) 1989-90

c) 1990-91 d) 1991-92

60. Rajiv Gandhi National Drinking Water Mission was launched in the year:

a) 1991 b) 1987

c) 1988 d) 1990

61. The 'Community-based rural drinking water supply' programme was:

a) Swajaldhara b) Jaldhara

c) Neerdhara d) Samritdhara

62. In which programme free and compulsory education to all children in the age group of 6-14 years is given:

a) Sarva Shiksha Abhiyan

b) Kasturba Gandhi Balika Vidyalaya

c) National Literacy Mission

d) National adult education programme

63. Sarva Siksha Abiyan was came in existence in the year:

a) 1994 b) 2001

c) 2006 d) 2004

64. Nehru Yuva Kendras was started in the year:

a) 1962 **b) 1972**

c) 1982 d) 1992

65. In which year 'Centrally sponsored Rural Sanitation programme' was launched:

    a) 1975 **b) 1986**

    c) 1991 d) 1996

66. First cooperative society was started in the year:

    a) 1904 b) 1912

    c) 1928 d) 1935

67. Commercial banks was nationalized in the year:

    a) 1965 b) 1967

    c) 1969 d) 1971

68. Regional Rural Bank was set up in the year:

    a) 1935 b) 1969

    c) 1972 **d) 1975**

69. Sponsoring agency for 'Vikas Volunteer Vahini' scheme is:

    a) DRDA b) RRB's

    c) NGO's **d) NABARD**

70. Kissan Credit Card was firstly launched by:

    a) Punjab National Bank **b) State Bank of India**

    c) Cooperation Bank d) Indian Bank

71. The total number of bank linked with SHG in India:

    a) 11.25 lakh **b) 48.51 lakh**

    c) 63.25 lakh d) 34.25 lakh

72. The weather insurance scheme of government of India was:

    a) Ganga Beema **b) Varsha Beema**

    c) Dhan Phuhar Beema d) Barsat Beema

73. The year in which, ICAR was established:

    a) 21rd March 1929 b) 29rd April 1929

    c) 23rd May 1929 d) 28rd June 1929

74. Among them which recommended the establishment of ICAR:

    a) Ford foundation **b) Royal commission**

    c) World Bank d) European union

75. The president of ICAR is:

a) Prime Minister
b) Agricultural Minister
c) President of India
d) Finance Minister

76. KVK was established by:

a) B R Mehta Committee
b) Kothari Committee
c) M S Swaminathan Committee
d) M S Mehta Committee

77. The year in which first KVK was established:

a) 1973
b) 1974
c) 1975
d) 1976

78. Where the first KVK was established:

a) Calcutta
b) Ranchi
c) Coimbatore
d) Pondichery

79. The funding agency of KVK is:

a) ICAR
b) SAU
c) State Government
d) Central Government

80. From which country 'Journal of extension' is published:

a) UK
b) Germany
c) USA
d) Holand

81. The 4 H Club was founded by:

a) O H Benson
b) A B Graham
c) Seman A Knapp
d) A B Ross

82. The motto of 4 H Club is:

a) The poorer rich
b) To make better best
c) Adult motivated
d) Child educated

83. PRADAN stands for:

a) Private Assistance for Development Action
b) Professional Assistance for Development Action
c) Professional Action of Development Authority
d) None

84. PRADAN is an:

a) Community group
b) SHGs
c) NGO
d) None

85. MYRADA stands for:
    a) Mysore Rehabilitation and Development Agency
    b) Mysore Research and Development Agency
    c) Mysore Resettlement and Development Agency
    d) Mysore Resettlement and Development Authority
86. Full form of SRISTI is:
    a) Summit for Research and Initiatives for Sustainable Technologies and Institutions
    b) Society for Research and Initiatives for Sustainable Technologies and Institutions
    c) Society for Rehabilitation and Initiatives for Sustainable Technologies and Institutions
    d) None
87. SRISTI is an NGO setup in the year:
    a) 1996 b) 1999
    c) 1993 d) 1998
88. In How many language, SRISTI newsletter published:
    a) 11 b) 06
    c) 09 d) 03
89. SRISTI newsletter in English language published by which name:
    a) SHAKTI **b) HONEYBEE**
    c) SAMPOORNA d) None
90. AASTIIK stands for:
    a) Academy for Augmenting Sustainable Technological Inventions, Innovations, and Traditional Knowledge.
    b) Assembly for Augmenting Sustainable Technological Inventions, Innovations, and Traditional Knowledge.
    c) Academy for Augmenting Sustainable Technological Interventions, Innovations, and Traditional Knowledge.
    d) None
91. AASTIIK began as an independent programme in the year:
    a) 2005 b) 2007
    c) 2011 d) 2008

92. Federation of Indian Women Entrepreneurs is a National-level organization, founded in

    a) 1993 b) 1990

    c) 1996 d) 1998

93. Consortium of Women Entrepreneurs registered as:

    a) NGO

    b) Civil society, Nonprofit Organization

    c) SHGs

    d) Civil society, Profit Organization

94. Association of Lady Entrepreneurs of Andhra Pradesh registered under:

    a) Society Act, 1966 **b) Company's Act 1956**

    c) Both d) None

95. In TiE Stree Shakti 'TiE' indicates:

    a) The Indus Entrepreneurs b) The Indian Entrepreneurs

    c) The Indian Entrepreneurship d) The Indus Entrepreneurship

96. In which amendments, constitution of India have provided reservation of seats in the local bodies of Panchayats and Municipalities for women:

    a) 73rd and 74th b) 72th

    c) Both d) None

97. The National Commission for Women was set in the year:

    a) 1992 b) 1999

    c) 2001 d) 1995

98. Who is the first woman prime minister of India?

    a) Smt. Indira Gandhi b) Smt. Meera Nayer

    c) Kiran Bedi d) None

99. Who is the first woman President of India?

    a) Smt. Indira Gandhi b) Smt. Meera Nayer

    c) Smt. Pratibha Devisingh Patil d) None

100. Women's Empowerment International Foundation (WEIF) is working on:

    a) Central and South Asia b) America

    b) Both d) None

101. Who is the First Woman Director of Directorate of Research on Women in Agriculture (DRWA)?

a) Dr. Hema Pandey b) Dr. B Meena kumari

c) Dr. Krishna Srinath d) Dr. Krishna Tirath

102. Who is the first chairperson of Bhartiya Mahila Bank?

a) Ms. Kavita Krishnamoorthy

b) Ms. Usha Ananthasubramanian

c) Ms. Usha Thorat

d) Ms. Arundhati Bhattacharya

103. Who is the chairperson of National Commission on Women (2019)?

a) Mrs. Rekha Sharma b) Mrs. Kiran Bedi

c) Mrs. Sucheta Bhardwaj d) Mrs. Nirula Sindhe

104. If women have same access as men to land, technology, financial services, education and markets, it would be possible to reduce the number of hungry people in the world by________

a) 10 to 15 million **b) 100 to 150 million**

c) 500 to 1000 million d) 1 to 2 million

105. Reservation of seats for women in panchayati raj institutions seeks to ensure ______

a) Economic empowerment of women

b) Superiority of women over men

c) Participation of women in public life

d) Strengthening women as an individual

106. The government has set up a committee of experts to prepare the blueprint of India's first Women's bank. The committee is headed by whom?

a) O P Bhatt b) B A Prabhaka

c) Pratip Chaudhuri **d) M B N Rao**

107. International Women's Development Agency (IWDA) established in 1985 is situated at which of the following places?

a) Paris b) New York

c) Melbourne d) New Delhi

108. The term "gender typing" with regard to work refers to __________
   a) Adopting biased hiring practices for men and women
   b) Labeling certain management techniques as "male" or "female"
   c) Seeing certain jobs as either masculine or feminine and then become dominated by one gender or the other
   d) Discriminating men and women for similar workplace achievements

109. First Indian woman to win the Booker Prize
   a) Arundhati Roy
   b) Nirja Mishra
   c) Vandana Shiva
   d) None of the above

110. Which agencies are responsible for creating gender difference in society?
   a) Family
   b) Social system
   c) Religion
   d) All the above

111. ________ is a strategy for incorporating women's as well as men's concerns and experiences in policies and programmers at all levels so that it will benefit both of them equally and prevent further inequality.
   a) Gender mainstreaming
   b) Gender Parity
   c) Gender prosperity
   d) All the above

112. Practical Gender Needs (PGN) are ________
   a) Short term
   b) Immediate
   c) Long term
   d) Both a & b

113. Strategic Gender Needs (SGN) are ________
   a) Short term
   b) Immediate
   c) Long term
   d) Both a & b

114. Practical Gender Needs (PGN) pertains to ________
   a) Day to day living conditions of women
   b) The position of women vis-à-vis men
   c) Both a & b
   d) None of these

115. Strategic Gender Needs (SGN) pertains to ________
   a) Day to day living conditions of women
   b) The position of women vis-à-vis men
   c) Both a & b
   d) None of these

116. Practical Gender Needs (PGN) includes ________
   a) Domestic income earning activities
   b) Housing and basic services
   c) Traditional employment opportunities
   d) All the above

117. Strategic Gender Needs (SGN) changes in _________
   a) The gender division of labour
   b) Shared domestic labour and child care
   c) Elimination of institutionalized forms of discrimination
   d) All the above

118. "Stree Shakti Puraskar" awards are named after which of the following eminent women?
   a) Rani Laxmi Bai & Mata Jijabai
   b) Rani Rudramma Devi & Kannagi
   c) Rani Gaidinliu Zeliang & Ahilya Bai Holkar
   d) All the above

119. Which country declares official National Holiday on 'Women's Day'?
   a) Russia
   b) India
   c) Canada
   d) Australia

120. What is the correct expansion of ICRW?
   a) International committee on Review of Women
   b) International Centre for the Research on Women
   c) International Conference on Research of Women
   d) International Council for Research on Women

121. Multiplier training by the trained farm women and organizing women into groups were the special component of _________
   a) TANWA programme
   b) WYTEP programme
   c) TEWA programme
   d) MAPWA programme

122. Farm Women Groups (FWGs), which were started with about 150 groups, gradually evolved into Self-Help Groups (SHGs) and later, some groups transformed in to "micro entrepreneurs' under which of the following programme?
   a) TANWA programme
   b) WYTEP programme
   c) TEWA programme
   d) MAPWA programme

123. Training & Extension for Women in Agriculture (TEWA) project of Odisha was started in which of the following years?

a) 1999
b) 2000
c) 1987
d) 1985

124. Madhya Pradesh Women in Agriculture (MAPWA) project was started in 8 districts of Madhya Pradesh in which of the following years?

a) 1993
b) 1989
c) 1999
d) 1995

125. First woman to become Deputy Chairperson of Rajya Sabha was ________

a) Violet Alva
b) Meera Nayar
c) Najma Heptullah
d) Margaret Alva

126. According to Global Gender Gap Index 2013-what is India's status out of 136 countrries?

a) 101
b) 105
c) 110
d) 99

127. The share of funds under ATMA scheme to be utilized to benefit the women farmers ________

a) 30%
b) 35%
c) 50%
d) 10%

128. Who is the first woman to receive Bharat Ratna Award?

a) Lata Mangeshkar
b) Sarojini Naidu
c) Indira Gandhi
d) Mother Teresa

129. In two way classification in Gender Budgeting Statement, Part B includes

a) Schemes with 100% allocation for women
b) Schemes with 30 to 99% allocation for women
c) Schemes with <30% allocation for women
d) All the above

130. The gender inequality more deep rooted in society because of ________

a) Traditional norms
b) Religious biliefs
c) Dominance men folks of the society
d) All the above

131. Harvard Analytic Framework developed by Harvard Institute of Development is also known as __________

a) Gender Roles Framework
b) Gender Planning Framework
c) Gender Analysis Matrix
d) Social Relations Framework

132. Caroline Moser Framework developed by Caroline Moser is also referred as _____

a) Gender Roles Framework
b) Gender Planning Framework
c) Gender Analysis Matrix
d) Social Relations Framework

133. Rani Parker Framework developed by Rani Parker is also called as ________

a) Gender Roles Framework
b) Gender Planning Framework
c) Gender Analysis Matrix
d) Social Relations Framework

134. Naila kabeer Framework is also called as __________

a) Gender Roles Framework
b) Gender Planning Framework
c) Gender Analysis Matrix
d) Social Relations Framework

135. Socio-Economic and Gender Analysis (SEAGA) is an approach elaborated by _____

a) FAO
b) UN women
c) UNDP
d) UNICEF

136. Women treated better and enjoyed more rights in which of the following period?

a) Vedic period
b) Post Vedic period
c) Both a & b
d) None of these

137. Extraordinarily talented women like Aditi, Lopamudra, Gargi, Maitreyi Ghosha and Appala contributed to which of the following period?

a) Rigveda
b) Samveda
b) Ayurveda
c) Yajurveda

138. Sankaracharya says if no shakti, then no ________

a) Shivam
b) Bhakti
b) Laxmi
c) Mukti

139. Who opined, Gods reside where women are respected?

a) Manu b) Sankaracharya

b) Shukracharya c) All the above

140. Legislation usually fixes a land ceiling for each ___________

a) Household b) Village

b) Society c) Subdivisin

141. Equal Remuneration Act of 1976 prohibited ___________

a) Sex discrimination b) Age discrimination

b) Skill based discrimination c) Colour based discrimination

142. When was the Family Court established?

a) 1984 b) 1999

c) 2001 d) 1994

143. The maternity Benefit Act, 1961 provides ___ for maternity leave with pay.

a) 25 weeks b) 20 weeks

b) 30 weeks c) 15 weeks

144. Who was the first male appointee to the National Commission for Women (NCW) in India?

a) Alok Nath b) Alok Kumar

b) Alok Rawat c) Alok Chatterjee

145. Which of the following is true about 'Gender Audit'?

a) Social Audit b) Quality Audit

b) Tool and a process **c) All of these**

146. Which SDG is related to gender equality?

a) SDG 5 b) SDG 7

b) SDG 8 c) SDG 10

147. Under one stop centre-scheme women can complain after ________ days

a) 7 days b) 3 days

b) 5 days c) 10 days

148. Women component plan was the concern in which plan?

a) 8th five year plan b) 5th five year plan

b) 9th five year plan c) 3rd five year plan

3

Rural Sociology (Theory)

- Father of sociology: Auguste Comte
- The term sociology was coined by Auguste Comte (1789-1875).
- Sociology word comes from two roots i.e. Latin word 'socius' meaning companion Greek word 'logos' meaning speech or reasoning.
- **Sociology:** Sociology is the scientific study of human social behavior and its origins, development, organizations, and institutions. It is a social science that uses various methods of empirical investigation and critical analysis to develop a body of knowledge about human social actions, social structure and functions.
- **Rural sociology:** The study of social life of peoples lives in rural areas.
- **Social organization:** The study of the various institutions, social groups, social stratification, social mobility, bureaucracy, ethnic groups and relations, and other similar subjects such as education, politics, religion, economy etc.
- **Social psychology:** The study of human nature as an outcome of group life, social attitudes, collective behavior, and personality formation.
- **Social change:** Alteration in the structure and function of society. It may be structural, functional and cultural changes and it deals with the study of such current problems in society such as juvenile delinquency, criminality, drug addiction, family conflicts, divorce, population problems etc.
- **Human ecology:** Deals with the nature and behavior of a given population and its relationships to the group's present social institutions.
- **Population or demography:** The study of population number, composition, change, and quality as they influence the economic, political, and social system.
- **Applied sociology:** The findings of pure sociological research in various fields such as criminology, social work, community development, education, industrial relations, marriage, ethnic relations, family counseling, problems of daily life etc utilized in this.

Family: A family word came from root latin- familia which means 'group of people' affiliated by consanguinity (recognized by birth), affinity (by marriage), or co-residence/shared consumption. OR

According to Maciver & Page 1977 "The family is a group defined by a sex relationship sufficiently precise and enduring to provide for the procreation and upbringing of children". Family is the basic unit of society as well basic unit of Extension work. Family is also an important economic unit studied in family economics.

Genealogy: This is a field which aims to trace family lineages through history.

Nuclear Family: Merriam-Webster used the term in year 1947.

Classification of family

1. According to blood relation living together:
   a) Nuclear: Husband, wife & unmarried children.
   b) Joint: Number of generation living together.
2. According to stage of life:
   a) Family of origin/orientation: Person is born and gets initial orientation.
   b) Family of procreation: Family which setup after marriage.
3. According to numbers of husbands and wife:
   a) Polygynous: Man marries more than one women.
   b) Polyandrous: Woman marries more than one men.
4. According to rule of lineage:
   a) Patri lineal: Determination of descent and property take place with the male line.
   b) Matri lineal: Determination of descent and property take place with the female line.
5. According to rule of residence:
   a) Patri local: Married couple and its offspring put up with the husband's family.
   b) Matri local: Married couple and its offspring put up with the mother's family.

Group

According to Chitambar "Social group is a unit of two or more people in reciprocal (to and fro) interaction and in communication with each other".

According to Maclever "Social group as a collection of human beings who enter into distinctive social relationships with one another".

Types of Social Groups

a) Based on Type of relationship	
Primary Group	**Secondary Group**
E.g. family, friendship group, play group etc	E.g. political groups, labour unions, trade unions, employees associations etc
b) Based on Mode of organization and functioning	
Formal Group	**Informal Group**
E.g. Labour union, village council, students union etc	E.g. family, friendship group, play group
c) Based on Structure and type of membership	
Voluntary Group	**Involuntary Group**
E.g. friendship group, play group etc	E.g. family, neighborhood, community
D. Based on Social class	
Horizontal Group	**Vertical Group**
E.g. caste	E.g. race, nation etc
E. Based on Personal feelings	
In Group	Out Group
E.g. my family, my class, my church etc.	E.g. their family, their class, their church
F. Based on groups Size	
Small Group	**Large Group**
E.g. family, play group etc.	E.g. political group, labour union etc

Delegate group: The members of this group are representative and chosen by groups. E.g. Gram Panchayat, U.N.O., etc.

Community: According to Chitambar 1990 "Community refers to groups of mutually dependent people, living in a more or less compact geographical area, having a sense of belonging and sharing common values, norms and some common interest and acting collectively in an organized manner to satisfy their chief needs through a common set of organizations and institutions".

Social System: Defined as a set of interrelated units that are engaged in joint problem solving to accomplish a common goal.

Social Mobility: Movement of individuals or groups from one stratum of society to another. In a social system three type of mobility can be seen i.e. Vertical, Horizontal and Geographical social mobility.

Social stratification: According to Lundberg 1968 is the division of population into two or more layers, each of which is relatively homogeneous and between which there are differences in privileges (opportunities), restrictions, rewards and obligations.

Forms of social stratification

1. Caste system
2. Class system

Class: Number of individuals in the same society whose status is similar.

Caste: The term 'Caste' was derived from the Portuguese word 'casta' meaning lineage or race. A caste is a social category whose members are assigned a permanent status within a given social hierarchy and whose contacts are restricted accordingly. Caste is a closed class. As compared to class, the caste is the most rigid, clearly graded type of social stratification and often referred to as one of the extreme forms of closed class systems.

Social interaction: According to Sutherland 1961 "Social Interaction is dynamic interplay of forces in which contacts between persons and groups results in a modification of the attitudes and behavior of the participants. E.g. Cooperation, accommodation, assimilation (positive interaction) and competition, conflict (negative interaction).

Social processes: Social processes refer to repetitive forms of behaviour which are commonly found in social life. E.g. Cooperation, competition, accommodation and assimilation

Social Institutions: According to Horton and Hunt 1976 "institution is an organized system of social relationship which embodies certain common values and procedures and meets certain basic needs of the society". The family is the most basic social institutions.

Culture: Culture is a way of life, mode of thinking, acting and feeling OR refers to totality of what is learned by individuals as members of the society. According to Ralph Linton culture may be the thought in terms of three different orders:

Two overt (which can be seen):

1. Material products of industry, implements, tools etc
2. Overt behavioral patterns like customs, folkways etc

One covert (which cannot be seen or hidden):

3. Psychological like attitudes, values etc. held by the individuals or groups

Social control: It is the pattern of influence the society exerts on individuals and groups to maintain order and establish rules in the society. Folkways, mores, norms and laws are the various form of social control.

Ethnocentrism: Ethnocentrism refers to that ours is the best of all cultures and the way we do things is the right way to do them.

Customs: The sociologists have used various terms in order to classify various human acts of behavior. If these various types of human behavior are organized, they are called customs.

Folkways: Folkways are the customary ways of behaving in society, in which society exerts some force for conformity

Mores: Mores are the plural of Latin word 'More'. The mores are the customs or patterns of behavior, which are regarded by members of social system as vital and essential to the welfare of the group. Generally the term 'more' is used for the positive action or things that ought to be done. Mores are the socially acceptable ways of behavior that do involve moral standards and violation of more may result in severe social action, such as ostracism (exclusion of individual or family from the village or society).

Taboos: Taboo means forbid. It refers to the prohibitions of the types of behavior because of some magical, supernatural or religious sanction. The term 'taboo' is used for the negative action and for the things that one ought not to do.

Rituals: Ritual is prescribed form of behavior for certain occasions and certain actions are designated in prescribed manner.

Values: Value is defined as anything desired or chosen by someone.

Prejudice: Judgment before due examination and consideration of fact and based on certain assumptions generally lead to the formation of prejudice. It is normally negative and difficult to reverse.

Stereotype: They are the fixed image formed in one's mind about people, practices or other social phenomena on the basis of experience, attitudes, values etc.

Attitude: It is a positive or negative feeling one has towards any psychological object. Attitudes are based up on values and attitudes influence our action positively or negatively. Hence they are covert and not overt.

Values and Norms: Values are the attitudes, held by the individuals, groups or society as a whole, as to whether material or non-material objects are good, bad, desirable or undesirable. Norms are the accepted and approved forms of behavior that are based on and consistent with dominant social values in society.

Social control: According to Mannheim "Social Control is the sum of those methods by which a society tries to influence human behavior to maintain a given order".

Leader: A leader is "a person who influences a group of people towards the achievement of a goal".

Type of leader

1. Autocratic leader: Autocratic leader is also known as authoritarian leader. He operates as if he cannot trust people. He thinks his subordinates are never doing what they should do; that the employee is paid to work and therefore must work.
2. Democratic leader: He shares with the group members the decision making and planning of activities. The participation of all members is encouraged. He works to develop a feeling of responsibility on the part of every member of the group.
3. Laissez-faire leader: He believes that if you leave workers alone, the work will be done. He seems to have no confidence in himself. If at all possible he puts off decision-making. He tends to withdraw from the work group.

Note: Rogers designated the Local leader as Opinion leaders.

Psychology: Father of Psychology: Sigmund Freud.

According to Daniel "Psychology is the scientific study of the behavior of an individual"

According to Murthy "Psychology is the science that studies the responses which living individuals make to their environment".

Intelligence

According to Thorndike "Intelligence is the ability of an individual to make profitable use of past experience"

According to Grprett "Intelligence is the ability demanded in the solution of problems, which require the comprehension, and the use of symbols"

$$\text{I.Q.} = \frac{\text{Mental age of an individual (MA)}}{\text{Chronological age of an Individual (CA)}} \quad 100$$

I.Q. Level	Character
Below 20	Idiot
20 to below 70	Feeble minded
70 to 90	Dull

90 to 110	Average or Normal
110 to 120	Superior
120 to 140	Very superior
140 to 200	Genius
Above 200	Supreme genius

Personality: The word Personality originated from the Latin word 'Personare' which used to mean the voice of an actor speaking through a mask.

According to J. B.Watson 1919 "Personality is the result of what we start with and what we have lived through. It is the reaction mass as a whole".

Perception

According to Kuppu Swamy "is the process by which impressions, opinions and feelings about an object is formed by means of a sensory operation"

Frustration: It is a psychological and physiological response, which occurs with the individual due to some obstacle in his goal directed behaviour OR Frustration is a condition where in the goal directed behaviour of an individual is blocked or thwarted.

Emotion: The word Emotion is originated from the Latin word 'emovere'.

'e'	:	means out and
'Movere'	:	means to move.

Emotion generally relates with the psychological phenomena like fear, anger, grief, love, affection, pleasure etc. Emotion is a stirred up state of feeling that is the way it appears to the individual himself.

Motivation: Defined as goal seeking or goal directed behavior or activity.

Teaching: The process of providing situations in which learning takes place; in other words, arranging situations in which the things to be learnt are brought the attention of the learners, their interest is developed, desire aroused, conviction created, action promoted and satisfaction ensured.

Learning: Learning is a process of progressive behavior adaptation through self activity.

Multiple Choice Questions

1. Who is the father of sociology?

   a) Auguste Comte b) Herbert Spenser

   c) Emile Durkheim d) Max Weber

2. Main concept of sociology is:
   a) Family b) Community
   c) Village **d) Society**
3. Basic unit of society:
   a) Neighborhood b) Group
   c) Family d) Community
4. Origin of society was occurs due to:
   a) Force b) God's will
   c) Evolution d) Individual
5. In sociology, term society is used for:
   a) Persons living in an area
   b) Persons professing the same religion
   c) System of social relationships
   d) Organized relations between individuals
6. Organization of peoples for a particular purpose is called:
   a) Community **b) Association**
   c) Society d) Institution
7. The term 'bourgeoise' (type of society) was given by:
   a) Max weber **b) Karl marx**
   c) Hegel d) T H Green
8. Acculturation, social integration and assimilation comes under:
   a) Steps in social movement b) Steps in assimilation process
   c) Forms of accommodation d) Anti-competitive behaviour
9. Class is an social stratification based on:
   a) Co-operation b) Competition
   c) Individual capability **d) Economic condition**
10. Condition which nit necessary for a group:
   a) Common locality b) Common ancestry
   c) Common interests **d) Common food habits**
11. A cooperative society comes under:
   a) Formal group b) Informal group
   c) Primary group d) Territory group

12. The critical form of social stratification is:
    a) Primary and secondary group b) Rich and Poor
    c) Male and female group **d) Class and caste**
13. Which type of co-operation among family members is:
    a) Nonformal b) Voluntary
    c) Formal **d) Informal**
14. The main character of institutions are:
    a) Social relationship b) Approval by culture
    c) Competition **d) Both a and b**
15. A codified system of mores is:
    a) Norms **b) Laws**
    c) Customs d) Values
16. Take the odd man out indicates:
    a) Competition b) Accommodation
    c) Assimilation d) Cooperation
17. The concept of gemeinschaft and gesellschaft was given by:
    a) E M Rogers **b) F Tonnies**
    c) Max Webber d) Keynees
18. When the peoples move outside the society iiiis called:
    a) Immigration **b) Emigration**
    c) Migration d) All of the above
19. By totalitarian society, which type of institution are arise:
    a) Catalystic b) Voluntary
    c) Prescriptive d) Lazerfaire
20. Co-operation is a process of:
    a) Conjunction process b) Disjunctive process
    c) Dissociation process d) Disintegrative process
21. Among them which of the following geographically demarcated:
    a) Society **b) Community**
    c) Group d) None of the above
22. Value judgment is high in:
    a) Folkways b) Rituals
    c) Mores d) Taboos

23. Study of population is known as:
    a) Dermatology
    b) Demagogy
    c) Anthropology
    d) Demography
24. In a society, social stratification occurs due to:
    a) Differentiation and evaluation
    b) Differentiation and classification
    c) Classification and satisfaction
    d) Classification and tabulation
25. Kinship group is also known as:
    a) Voluntary group
    b) Involuntary group
    c) Delegate group
    d) Associated group
26. Sociometry methods were given by:
    a) Schramm
    b) Rogers
    c) Moreno
    d) Shoemaker
27. Who is the biologically inbred group of people?
    a) Caste
    b) Race
    c) Class
    d) Religion
28. Family is an example of:
    a) Primary group
    b) Secondary group
    c) Pertaining group
    d) None of the above
29. Caste is a class gained by:
    a) Status
    b) Birth
    c) Hard work
    d) Social relationship
30. Assimilation is a:
    a) Social process
    b) Political process
    c) Historical process
    d) Psychological process
31. In a social system, socialization is a process which:
    a) Ceases at maturity
    b) Continues throughout the life
    c) Influences society
    d) None of the above

32. The term 'Barter' indicates the exchange of:
    a) Service of service
    b) Goods for service
    c) Goods for goods
    d) All of the above
33. Which is not an institution:
    a) Marriage
    b) Theatre
    c) Inheritance
    d) Property
34. Sex, age and caste are described the:
    a) Achieved status
    b) Ascribed status
    c) Pre-set status
    d) Status image
35. The primary group mainly recognized by:
    a) Lack of identification
    b) Affection and love
    c) Identification of interest
    d) Continuous contact
36. Among them is not a character of in-group:
    a) Jealousy and competition
    b) Intimate relationship
    c) Sense of achievement
    d) Element of sympathy
37. Group life determine by:
    a) We feeling
    b) Feeling of unity
    c) Common purpose
    d) All of the above
38. The word family originate from root:
    a) Greek
    b) Roman
    c) Latin
    d) British
39. Marriage by mutual affection is:
    a) Gandharva
    b) Rakshasa
    c) Daiva
    d) Asura
40. The most intensive class struggle is found in:
    a) Capitalistic society
    b) Primitive society
    c) Feudal society
    d) Socialistic society
41. Social stratification indicates:
    a) Social justice
    b) Social inequality
    c) Social injustice
    d) Social equality
42. Culture indicates:
    a) Social heritage
    b) Social etiquette
    c) Religious and moral attitudes
    d) None of the above

43. Factors responsible for cohesion of rural society are:
    a) Common traits
    b) Common objectives
    c) Similar experiences
    d) All of the above
44. Positive aspect of bureaucracy is:
    a) Red tapism
    b) Mechanical adherence to rules
    c) Objective orientation
    d) Indifference to individual feelings
45. Among them conformity to norms depends on:
    a) Effectiveness of sanctions
    b) Ideal character of the norms
    c) Cohesiveness of the group
    d) Sense of cooperation in the group
46. In which type of society role conflict is a major phenomenon:
    a) Primitive society
    b) Agarian society
    c) Tribal society
    d) Industrial society
47. Type of family on the basis of blood relationships is known as:
    a) Conjugal or consanguineous
    b) Matrilineal or patrilineal
    c) Nuclear or extended
    d) Matrilocal or patrilocal
48. Socialization is a process of converting a biological organization into:
    a) Human being
    b) Super human being
    c) Modern man
    d) A civilized person
49. To borrowing cultural items from another culture is the process of:
    a) Acculturation
    b) Socialization
    c) Unculturation
    d) Acquisition
50. The concept of primary groups was given by:
    a) Cooley
    b) Durkheim
    c) Ginsberg
    d) Sorokin
51. A man marrying more than one women is termed as:
    a) Polygyny
    b) Polyandry
    c) Hypergamy
    d) Polygamy
52. In social change, changes indicates in:
    a) Values system
    b) Dress
    c) Food habits
    d) Habitation patterns

53. Example of primary group is:

a) Family b) Students of sociology

c) Social workers d) All of the above

54. Social change means a change in:

a) Occupation b) Administration

c) Standards of living **d) Social structure**

55. Concept of social evaluation first given by:

a) Herbert spencer b) August comte

c) Emile Durkheim d) Lester C Ward

56. Who among the following actually imitates the action in the community:

a) Professional leader b) Assured leader

c) Operational leader d) Popular leader

57. Sociometry method was given by:

a) Fisher b) Osgood

c) Mereno d) William Stephenson

58. In sociometry methods two or more individual mutually choose each other is called:

a) Fraternal twins b) Identical twins

c) Semantic twins d) Clique

59. Social vacuum theory was given by:

a) Sorokin b) Mc Gregor

c) Curt Lewin d) Gilford Pinchot

60. The president of panchayat on a village is an:

a) Lay leader b) Action leader

c) Professional leader d) None

61. Psychology studies the:

a) Human behaviour b) Animal behaviour

c) Both a & b d) None of the above

62. Origination of psychology is form:

a) Physiology **b) Philosophy**

c) Biology d) Sociology

63. For human beings, audible range of sound waves is between:
    a) 20 and 30000 Hertz
    b) 20 and 2000 Hertz
    c) 20 and 20000 Hertz
    d) 10 and 30000 Hertz
64. Classical experiment on conditioning was done by scientist:
    a) Pavlov
    b) Bolles
    c) Kamin
    d) Menzel
65. Conditional stimulus is known as:
    a) Neutral stimulus
    b) Specific stimulus
    c) General stimulus
    d) Applied stimulus
66. In Pavlov's experiment, conditional stimulus was:
    a) Light
    b) Bell
    c) Food
    d) Saliva
67. Among which of the following branch of psychology deals with cognitive process is:
    a) Developmental psychology
    b) Clinical psychology
    c) Experimental psychology
    d) Cognitive psychology
68. Coded information put into memory is:
    a) Retrieval
    b) Encoding
    c) Storage
    d) None of the above
69. Dissonance and cognition respectively means:
    a) Thoughts and clashing
    b) Clashing and thoughts
    c) Thoughts and balancing
    d) All of the above
70. Storage of information from past experience related to:
    a) Information
    b) Motivation
    c) Memory
    d) Cognition
71. Formula for IO calculation was:
    a) Mental age + chronological age/100
    b) 100/Mental age x chronological age
    c) Mental age/Chronological age x 100
    d) Chronological age/Mental age x 100
72. A person obtaining a score of 120 in the IQ test is considered to be:
    a) Average intelligence
    b) Superior
    c) Low Intelligence
    d) None of the above

73. A person with mental ability far below average is termed as:
a) Borderline
b) Average
c) Mentally retarded
d) Dull normal

74. Biological needs are maintain the:
a) Thermostat
b) Homeostasis
c) Incentive value
d) Need

75. Curiosity is an example of:
a) Primary motive
b) Secondary motive
c) Stimulus motive
d) All of the above

76. Achievement is an example of:
a) Primary motive
b) Secondary motive
c) Stimulus motive
d) All of the above

77. Thirst is an example of:
a) Primary motive
b) Secondary motive
c) Stimulus motive
d) All of the above

78. A desire to meet some standard of excellence in a work is called:
a) Need for security
b) Need for power
c) Need for achievement
d) Need for self-actualization

79. In the hierarchy of Maslow, need that occupies the bottom most position is:
a) Psychological need
b) Biological need
c) Physiological need
d) Physical need

80. The basic speech sounds are:
a) Phonemes
b) Morphemes
c) Heuristics
d) Programming

81. Thematic apperception test was given by:
a) Cattel
b) Malvin Lee
c) Philip Zimbardo
d) Henry Murray

82. If correlation coefficient is zero indicates:
a) Nonexistent relationship
b) Existent relationship
c) Neutral relationship
d) None of the above

83. Which among the following is the strongest need:
    a) Hunger **b) Thirst**
    c) Safety d) Sex
84. Ability to attain a desired goal is:
    a) Skill b) Attitude
    c) Competency d) Motivation
85. The character of achievement motivation is:
    a) Sense of excellence b) Sense of unique achievement
    c) Sense of long germ involvement **d) All of the above**
86. Thanophobia is a fear of:
    a) Fear of fire **b) Fear of death**
    b) Fear of blood d) Fear of animals
87. Autophobia is a fear of:
    a) Fear of water b) Fear of men
    c) Fear of food **d) Fear of being alone**
88. In a role play, respondent play his own role known as:
    a) Self-drama **b) Psycho-drama**
    c) Auto-drama d) None of the above
89. The book entitled 'I'm OK and You're Ok is written by:
    a) Thomas and Amy Harris b) Eric Berne
    c) Thomas Pele d) Paul Stephenson
90. Among which of the country 'Farmers Field School's' are originated:
    a) Philippines b) Bangladesh
    c) Srilanka **d) Indonesia**
91. Among which of the country 'Media forms' were originated:
    a) Canada b) USA
    c) India d) Russia
92. The focus of trainer in training sessions is essentially on:
    a) Learner b) Subject matter
    c) Learning climate d) Learning outcome
93. The outcome of training rest solely on:
    a) Learner b) Trainer
    c) Institution d) Both a and b

94. Human beings continue to learn till:
    a) Adulthood
    b) Childhood
    c) College days
    d) Life end
95. The number of trainees attending an open learning programme will be:
    a) Very small
    b) Small
    c) Medium
    d) Large
96. Krishi Vigyan Kendra is come under:
    a) Trainers Training Center
    b) Extension Workers Training Center
    c) Vocational Training Center
    d) Leadership Training Center
97. The apex organization for offering training for extension professionals:
    a) MANAGE
    b) NAARM
    c) NCAP
    d) NIRD
98. Normal pace of delivery of word in a lecture method is:
    a) 120 words per minute
    b) 160 words per minute
    c) 200 words per minute
    d) 240 words per minute
99. Use of psychological tools in training is otherwise known is:
    a) Instrumental exercise
    b) Stimulation exercise
    c) Sensitivity exercise
    d) Laboratory exercise
100. Among them which training method suits to all situation:
    a) Simulation method
    b) Lecture method
    c) Sensitivity method
    d) None of the above
101. Simulation method does not include:
    a) In basket exercise
    b) Role play
    c) Business game
    d) Brainstorming
102. Which type of leadership is a negative, pessimistic or discouraging approach?
    a) Leader centered
    b) Follower centered
    c) Public centered
    d) All of these
103. The concept of Sanskritization was given by?
    a) M N Srinivas
    b) Emile Durkheim
    c) Iravati Karve
    d) G.S Ghuriya

104. The term 'social fact' is related with the work of which of the following sociologist?

a) **Emile Durkheim**
b) Auguste comte
c) Sigmund Freud
d) Simon

105. The concept of 'Sociometry' was given by__________

a) Bogardus
b) Levi Strauss
c) J.L Moreno
d) F Tonnies

106. __________ is the practice of comparing other cultural practices with those of noe's own and automatically finding those other cultural practices to be inferior.

a) Xenocentrism
b) Egocentrism
c) Ethnocentrism
d) Cultural Lag

107. Which is the correct sequence of social evolution according to L.H Morgan?

a) Barbarism – Civilization – Savagery
b) Civilization – Savagery – Barbarism
c) Savagery – Barbarism – Civilization
d) Barbarism – Savagery – Civilization

108. The broad meaning of Sanskritization is:

a) Horizontal mobility
b) Structural mobility
c) Parochialisation
d) Cultural upward mobility

109. Working through 'socio – metric Stars' is suggested in __________

a) Progressive village
b) Non – progressive village
c) Both a & b
d) None of these

110. What is the correct term for becoming aware or knowing something?

a) Conation
b) Connotation
c) Cognition
d) Foresight

111. What is the unique, integrated and organized system of all behavior of a person?

a) Attitude
b) Personality
c) Leadership
d) Living standard

112. Who are the people neither introvert nor extrovert?

a) Assertive
b) Aggressive
c) Ambiverts
d) Shy person

113. First laboratory of psychology was established by whom?
   a) Adam Smith
   b) William Ouchi
   c) William Wundt
   d) Stephen Gerard
114. The word 'Introspection' is made up which language?
   a) German
   b) Greek
   c) Roman
   d) Latin
115. In simple word 'looking outside oneself' is denoted as-
   a) Supervision
   b) Emancipation
   c) Observation
   d) Systematization
116. Which one of the following is true about Perception?
   a) Mental Process
   b) Physical effort
   c) Feeling about mind
   d) Physical sensation
117. Who gave the Psychogenic Needs and Viscerogenic Needs concept?
   a) Henry Murray
   b) Abraham Maslow
   c) Oladele, J.O.
   d) Thorndike
118. Who among the following proposed the concept of 'General Intelligence'?
   a) Armando Colaso
   b) Victor Hugo
   c) Charles Spearman
   d) Dario Foe
119. 'General intelligence' is also known as which factor?
   a) A
   b) B
   c) F
   d) G
120. Who among the following has given the first three principles of learning?
   a) Simon
   b) George Kimble
   c) Readiness
   d) Edward Thorndike
121. The Full form of ESB is ___________
   a) Erythrocyte Sedimentation Bar
   b) Electrical Stimulation of the Brain
   c) Electronic Stimulation of Behaviour
   d) Emotional Standard Behaviour
122. Which among the following theories say 'emotions and arousal occur at the same time'?
   a) James – Biggs theory
   b) James – Lange theory
   c) Lewis Terman theory
   d) Cannon – Bard theory

123. Which theory refers 'emotion is the result of arousal'?

a) James – Biggs theory
b) James – Lange theory
c) Lewis Terman theory
d) Cannon – Bard theory

124. 'Phrenology' is associated with which of the following?

a) Measurement of intelligence
b) Measurement of personality
c) Measurement of emotions
d) Measurement of memory

125. Discrimination towards single people is known as ___________

a) Dualism
b) Hatred
c) Singlism
d) None of these

126. When Stimuli or events increase the availability in memory, it is known as ___________

a) Priming
b) Stimulating
c) Increasing
d) Invigorating

127. Behavior patterns that are typical of particular species of animals is known as

a) Biology
b) Ethology
c) Structuralism
d) None of the above

128. Emotional or evaluative meaning of a word or concept is known as ___________

a) Denotative
b) Connotative meaning
c) Sensational meaning
d) Conceptual clarity

129. Which of the following is incorrectly matched ?

a) Sanguine – Cheerful
b) Melancholic – Depressed
c) Phlegmatic – Unexcitable
d) Cholesic – Cool tempered

130. Which one of the following is not coming under the ABC of Behaviour?

a) Antecedent
b) Behaviour
c) Consequence
d) Affection

131. Village become urban when

a) 75% of population engaged in industry and service
b) Population is 5000 plus
c) Density is 400/ sq. km.
d) All the above

132. Who among the following is associated with 'Rural vacuum theory'

a) A T Mosher b) J P Leagan

c) Willium Ward d) Seaman Knapp

133. Social learning theory is given by

a) Albert Bandura b) Aaron Back

c) Sigmund Freud and William James d) Wolfgang Kohler

134. In groups willing to accept rules to work in group comes under

a) Performing **b) Norming**

c) Adjoining d) None of these

135. Which quarter of Johari window depicts the Highly Inter group Behaviour?

a) Arena b) Blind

c) Façade d) None of these

4

Research Methodology in Extension (Theory)

Research: The word research is composed of two words "Re" and "Search".

"Re" - 'Again or over again or a new'.

"Search" - 'To examine closely and carefully' or 'to test and try'.

Research is a careful, systematic, patient study and investigation in some field of knowledge undertaken to establish principles / policies.

Science: It is a systematized knowledge derived through observation, experimentation or any other systematic method or procedure in order to determine the nature of phenomenon or process being studied or the causes behind specific events.

Research Methodology: The study of conducting research is research methodology.

Pure/Basic/Fundamental Research: Gathering, knowledge is termed as 'pure' or 'basic' research. Just to gather knowledge in order to formulate or generalize theories or policies. Pure research involves developing and testing theories and hypotheses that are intellectually challenging to the researcher but may or may not have practical application at the present time or in the future.

Applied Research: To find an immediate solution for a practical problem. Eg: Social, economical and political trends prevailing in a country.

Descriptive Research: Survey or fact finding enquires of different kinds. It describes the actual prevailing state of various matter or affairs, existing at present. Eg: describes attitudes towards an issue.

Analytical Research: The researcher has to use information already existing and analyze these data to make a critical evaluation. Eg: document study / historical evidence.

Explanatory Research: Attempts to clarify or explain why and how, any particular research problem arises and can be solved.

Exploratory Research: Study undertakes to explore a new area or an unknown destination. Eg: pilot study or feasibility study.

Empirical Research (based on experiments or experience): The result obtained by adapting Empirical Research is considered to be most powerful i.e. evidence enclosed.

Correlation Research: Attempts to discover or establish the existence of a relationship/ interdependence between two or more variables/aspects of a situation.

Research Design: Research design is the conceptual structure within which research is conducted. It constitutes the blue print for the collection, measurement and analysis of data. A detailed blueprint used to guide a research study towards its objectives such as:

1. Random observation study
2. Exploratory/formulative study
3. Descriptive study design
4. Diagnostic /analytical study
5. Survey methods
6. Case study design
7. Experimental study design → Cross sectional design, A Projected design, Ex post facto design
8. Evaluative study design

Variables: A magnitude that varies is known as "variable". In other words, variables are the characteristics or conditions that can be observed, manipulated or controlled by the researcher.

Types of Variable

1. Dichotomous and Polytomous variable:
2. Independent and dependent
3. Continuous and Discrete
4. Qualitative and Quantitative
5. Stimulus and Response
6. Intervening and Extraneous

Dichotomous and Polytomous variable: Dichotomous variables which have only two value eg: male-female while polytomous variables have more than two value such as Knowledge, skill etc.

Dependent or Endogenous Variables: The change in one variable depends on the change in other variable, it is known as dependent or Endogenous variable. Demand --- Price (independent)

Independent or Exogenous Variable: The variable that causes the change in the dependent variable is known as independent or exogenous variable. Demand (Dependent) ---- Price

Continuous Variables: Values that can be expressed even in decimal poins are known as continuous variables. A continuous variable is one which is capable of being measured in any arbitrary degree of fineness or exactness. Eg: Age, height, intelligence, income, level of adoption etc. are some examples of continuous variable.

Discrete Variables: Those variables whose values can be determined by counting. These are not capable of being measured in any arbitrary degree of fineness or exactness because the variables contain a clear gap. Value that can be expressed only in integer values are called non continuous variables. Eg: No. of students in a class.

Qualitative Variables: Refer to those which cannot be manipulated after the research is started and which consist of categories that cannot be ordered in magnitude. It refers to quality or characteristic or attribute and hence known as attributive characters. Such as colour, race, sex, religion.

Quantitative Variables: Refer to those variables which are counted or that can be ordered in magnitude i.e. it may exist in greater or smaller amounts. Eg: age, income

Stimulus and Response Variables: Stimulus is the condition or manipulation created by the researcher so as to evoke a response in an organism. The general classes of things the researchers observe that relate to the environment, situation or condition of stimulation are referred to as stimulus variables. The stimulus variables, also known as action variables, may be items like a slide show, a field day, method demonstration etc. Any kind of behaviour of the respondent is called response or behavioural variable.

Extraneous Variable: The independent variable which is not directly related to the purpose of the study but affects the dependent variable is known as extraneous variables. The influence caused by the extraneous variable on the dependent value is technically known as "Experimental Error"

Confounded Relationship: The relationship between dependent and independent variable is said to be confounded by an extraneous variables.

Intervening Variables/Hidden Variable: At the time of research some variables are cannot be observed but they exist and effect the relationship between the independent and dependent variables is called intervening or hidden variable. Eg: anxiety, fatigue, motivation etc.

Theory: According to F.N.Kerlinger 'Theory is a set of interrelated concepts, definitions, and proposition that present systematic view of phenomena by specifying relations among variables with the purpose of explaining and predicting the phenomena'.

Hypothesis: Hypothesis- 'Hype' has its origin in Greek means 'less than / under / beneath'.

A Research hypothesis is a predictive statement, which is capable of being 'tested' using scientific methods, which involves independent and dependent valuables. Hypothesis is statements about the relation between two variables e.g.: relation between free supply of power and productivity of small livestock farms.

Null Hypothesis (HO): If both the variables are equally good, it is Null Hypothesis OR is the proposition or proposal that implies no effect on the phenomena.

Alternative Hypothesis: If one variable is considered superior to other or vice versa or if there is a difference, it is alternative hypothesis OR is the one predictive statement that implies some effect on the phenomena.

Type I and Type II Errors

i) **Type I** - Researcher may reject Ho, when it is true.

ii) **Type II** - Researcher may accept Ho, when it is false.

One tailed and Two tailed Tests: One tailed test rejects the null hypothesis when the sample mean is either greater or lower than the hypothesized value of the population mean. When the sample mean is both greater and lower than the hypothesized value of the population mean known as two tailed test.

Research Hypothesis: Hypothesis is tested by adopting scientific methods, it is known as Research Hypothesis.

Experimental hypothesis testing: Research in which the independent variable are manipulated. Control Groups: A group is exposed to usual conditions in an experimental hypothesis.

Experimental groups: The group is exposed to special or certain new conditions.

Treatments: The different conditions to which the experimental & control groups are subject to exposed.

A population: Population is defined as the collection of aggregate of individuals the totaling of observation. If a population is subdivided in some parts, then these parts are called sample units.

A finite population: A population is said to be finite if it consists of a limited number of elements so that it can be easily counted. For instance, the population of university students, the population of a village or city etc.

A infinite population: Infinite population is one whose size is unlimited and, therefore, its members cannot be counted. The populations of fishes in a river, the number of microbes in the soil etc. are examples of infinite population.

Sample: A sample may be defined as a selected number from a population to represent it OR is a segment of the population selected to represent the population as a whole. While sampling is defined as the process of selecting the sample from the population.

Sampling frame: The representation of the elements in a population from which the sample is selected is called sampling frame.

Sampling error: This difference between a sample estimate and its population value is referred to as sampling error.

Sampling Design: An art of obtaining a sample from a given population. The technique or procedure the researcher adopts for selecting items for the sample from the population or universe.

Probability sampling methods

- Simple random sampling
- Systematic random sampling
- Proportionate random sampling
- Stratified random sampling
- Cluster sampling
- Multi stage sampling

Non-Probability sampling methods are

- Quota sampling,
- Purposive sampling,
- Systematic sampling
- Convenient sampling.

Probability samples

- Simple random sample: Every member of the population has a known and equal chance of being selected.
- Stratified random sample: Population is divided into mutually exclusive groups such as age groups and random samples are drawn from each group.
- Cluster sample: The population is divided into mutually exclusive groups such as blocks, and the researcher draws a sample of the group to interview.

Non-probability samples

- Convenience sample: The researcher selects the easiest population members from which to obtain information.
- Judgment sample: The researcher uses his/her judgement to select population members who are good prospects for accurate information.
- Quota sample: The researcher finds and interviews a prescribed number of people in each of several categories.

Data Collection: Data Collection is the systematic gathering of information (data) for a particular purpose from various sources.

1. Primary Data: Refers to the data collected for the first time (Original data)
2. Secondary Data: Refers to the data that have already been collected and used earlier by somebody or some agency.Eg: Online database, Historical books etc.

Measurement: Measurement is assignment of numerals to objects/events according to rules.

Types of measurement scales: S.S.Stevens has classified the measurement scale in to four categories:

1. Nominal or classificatory scale
2. Ordinal or ranking scale
3. Interval scale
4. Ratio scale

1. The nominal or classificatory scale

A nominal scale enables the classification of individuals, objects or responses into subgroups based on a common/shared characteristic. A variable measured on a nominal scale may have one, two or more subcategories depending upon the extent of variation. This scale comes under lowest level of measurement.

2. Ordinal or ranking scale

Besides categorizing individuals, objects, responses or a property into subgroups on the basis of common characteristic, it ranks the subgroups in a certain order. They are arranged either in ascending or descending order according to the extent a subcategory reflects the magnitude of variation in the variable. For example, 'income' can be measured either quantitatively (in rupees and paisa) or qualitatively using subcategories 'above average', 'average' and 'below average'. 'Socioeconomic status' and 'attitude' are other variables that can be measured on ordinal scale.

3. Interval scale

An interval scale has all the characteristics of an ordinal scale. In addition, it uses a unit of measurement with an arbitrary starting and terminating points. For example, Celsius scale: 0°C to 100°C

4. Ratio scale

A ratio scale has all the properties of nominal, ordinal and interval scales plus its own property: the zero point of a ratio scale is fixed, which means it has a fixed starting point. Since the difference between intervals is always measured from a zero point, this scale can be used for mathematical operations. The measurement of variables like income, age, height and weight are examples of this scale. It has the highest level of measurement.

Questionnaire: Questionnaire can be defined as a group of questions designed to collect information from a specific subject. Usually send by post or mail.

Type of questionnaire

1. Standardized questionnaire
2. Fixed alternative questionnaire
3. Structured and Non- Structured questionnaire
5. Open and closed questionnaire
6. Mixed questionnaire
7. Pictorial questionnaire

Observation: Observation could be defined as the careful and systematic watching of facts as they occur in the course of nature.

Types of Observation

1. Structured – for descriptive research
2. Unstructured—for exploratory research
3. Participant and Non- participant observation
4. Control and Non-control observation
5. Disguised observation

Survey: Survey implies taking a general view of all cases or situations - a broad general consideration of a whole situation or field of circumstances.

Social Survey: According to Wells 'Social survey is a fact finding study, dealing chiefly with working class poverty and with the nature of problems of the community'. Approach most suited for gathering descriptive information.

1. Structured and Unstructured Surveys
2. General and specific survey
3. Regular and ad-hoc survey
4. Preliminary and final survey
5. Census and sample survey

Case Study: According to P.V.Young 'Case study is a methods of exploring and analyzing the life of a social unit be it that a person, a family, an institution, cultural group or even entire community'.

Interview Schedule: According to P.V.Young 'Interview is regarded as a systematic method by which a person enters more or less imaginatively into the life of a comparative stranger'.

Content Analysis: Analysis of the contents of an interview in order to identify the main themes that emerges from the responses given by the respondents.

Epistemological: A theory of knowledge that is particularly employed in to refer to a stance on what should pass as acceptable knowledge.

Ethnographic Studies: Ethnographic study is a method of field observation or observation of behaviour in natural setting.

Data: A collection of facts from which conclusion can be drawn or it is information that has been translated into a form i.e. more convenient to move or process.

Qualitative Data: Qualitative data is relating to distinctions based on quality, characteristic or attribute like color, race, sex, religion etc.

Quantitative Data: Quantitative data involving or relating to considerations of size and which can be measured like age, income, size of handholding, length of experience in farming.

Validity: Validity means to what extent the tools selected measures correctly and what it is supposed to measure. They are four types i.e. content (face validity), predictive, concurrent and construct validity.

Internal Validity: It demonstrates that certain conditions lead to other conditions and it requires use of multiple pieces of evidence from multiple sources to uncover convergent lines of enquiry. The researcher strives to establish a chain of evidence forward and backward.

External Validity: It reflects whether the findings or generalized or not beyond the immediate case or cases. The more variation in places, people and procedures, a case study can withstand and still yield the same findings means more external validity. Techniques such as cross-case examination and within case examination along with literature review helped to ensure external validity.

Reliability: Reliability means to what extent the tool is accurate and stable in measuring a given variable. According to Kerlinger 'Reliability is the ability of the measuring instrument to yield consistent results when applicable to the same sample'.

Note: "All that is reliable is valid but all those are valid is not reliable"

Thematic Apperception Test: The Thematic Apperception Test, or TAT, is a projective psychological test. Proponents of this technique assert that a person's responses reveal underlying motives, concerns, and the way they see the social world through the stories they make up about ambiguous pictures of people.

A Guttman Scale: Presents a number of items to which the person is requested to agree or not agree. This is typically done in a 'Yes/No' dichotomous format. In other words, on a guttman scale, items are arranged in an order so that an individual who agrees with a particular item also agrees with items of lower rank-order. well-known example of a guttman scale is the bogardus social distance scale and original beaufort wind force scale

Semantic Differential: It is a type of a rating scale designed to measure the connotative meaning of objects, events, and concepts. The connotations are used to derive the attitude towards the given object, event or concept. Osgood's semantic differential was an application to measure the semantics or meaning of words, particularly adjectives, and their referent concepts.

Likert Scale: The scale is named after its inventor, psychologist 'Rensis Likert'. Likert scaling is a bipolar scaling method, measuring either positive or negative response to a statement. Likert scales are arbitrary. The value assigned to a Likert item has no objective numerical basis, either in terms of measure theory or scale. It is a psychometric scale commonly involved in research that employs questionnaires, survey research etc. The format of a typical five-level Likert item is:

1. Strongly disagree
2. Disagree
3. Neither agree nor disagree
4. Agree
5. Strongly agree

Summated Rating Scale: Summated rating scales have been developed for assessment of people's characteristics, especially attitudes, beliefs, emotions, feelings, perceptions and personality in the social sciences. They are widely used in both social science applications and research. Summated rating scales are preferred to single rating scales because they tend to have better reliability. As with a single rating scale, each item in a summated rating scale yields a numerical score. One of the most popular is the likert scale. There are four characteristics of a summated rating scale.

1. The scale must contain two or more items.
2. The format of each item must be a rating scale.
3. The scale must be designed to assess an underlying construct that is quantifiable.
4. The items do not have correct answers.

Multiple Choice Questions

1. The word science literally indicates:
   a) To make knowledge b) To undertake social survey
   c) To do research d) To understand nature
2. The basic aim of science is:
   a) Theory b) Prediction
   c) Control d) Hypothesis
3. The word science had originated from:
   a) Latin b) Greek
   c) Persian d) Arabic
4. In science knowledge generation is done through:
   a) Interview **b) Research**
   c) Survey d) Case study

5. By using logical consideration, searching of truth is refers to:
   a) Philosophy b) Principle
   c) Scientific method d) None of the above

6. Which method mostly used by laggards for fixing of belief:
   a) Method of tenacity b) Method of authority
   c) Method of intuition d) Method of science

7. The characteristic feature "Agree with Reason" is refers:
   a) Method of tenacity b) Method of authority
   c) Method of intuition d) Method of science

8. When similar topic investigates repeatedly in which successive studies are influenced by preceding inquires are known as:
   a) Invisible college b) Research paradigm
   c) Research design **d) Research tradition**

9. The characteristic feature of scientific data is:
   a) Objective b) Measurable
   c) Predictable **d) All of the above**

10. Which of the following used for existing data, predict new observations and guide further research:
    a) Hypothesis **b) Theory**
    c) Definition d) Method

11. Intelligence define in terms of verbal reasoning, comprehension, analytical ability and numerical capacity called the:
    a) Operational definition **b) Constitutive definition**
    c) Conceptual definition d) Observational definition

12. Pure research means:
    a) Illuminates theory
    b) Uses highly technical language
    c) Shares multipurpose view
    d) Concentrate on the "Why" aspect

13. The difference between classical research and participatory research is:
    a) Control b) Content
    c) Methods d) Substance

14. A books consists of one aspect but in a detailed manner called:
    a) Abstracts b) Bulletin
    c) Monograph d) Encyclopedia
15. In a research thesis, summary and conclusion chapter is knows as:
    a) Epilogue b) Prologue
    c) Alphaloge d) Symatologue
16. Which country published the social science index?
    a) Netherlands **b) New York**
    c) Singapore d) Columbia
17. The Indian Council of Social Science Research is situated at:
    a) Bangalure **b) New Delhi**
    c) Meerut d) Karnal
18. The book entitled 'How to measure attitudes' written by:
    a) Fred, N. Kerlinger
    b) O.P. Dhama
    c) Nicodemus
    d) Miller Delbert, Henerson, Morris & Gibbon
19. Research design is an:
    a) Plan b) Strategy
    c) Structure **d) All**
20. Among them which research design indicates "After this, therefore caused by this":
    a) Experimental design b) Survey design
    c) Ex post facto design d) Explorative design
21. The term ex post facto was first given by:
    a) Chapin and Greenwood b) Festinger and Katz
    c) Goode and Hatt d) Campbell and Katona
22. In which research design, pilot study is conducted to gain more insight into the research problem is:
    a) Explorative design b) Diagnostic design
    c) Descriptive design d) Experimental design
23. In which research design hypothesis is essential:
    a) Experiment design b) Diagnostic design
    c) Descriptive design d) Both a and b

24. which research design used without hypothesis:

    a) Explorative design b) Diagnostic design
    c) Descriptive design d) Experimental design

25. The effect of independent variable on a single experiment was studied by:

    a) Factorial design b) CRD
    c) LSD d) RBD

26. Opinion polls is an example of:

    a) Explorative design b) Diagnostic design
    c) Descriptive design d) Experimental design

27. Among them to whom numerals or values are assigned:

    a) Mean b) Mode
    c) Variable d) Median

28. Variables are:

    a) Property b) Symbol
    c) Constructs **d) All of the above**

29. A quantity which can vary from one individual to another individual is known as:

    a) Mean b) Mode
    c) Variable d) Median

30. Independent variables are:

    a) Presumed cause b) Antecedent
    c) Manipulated **d) All**

31. Variable which cause the changes in the independent variable is:

    a) Dependent variable b) Independent variable
    c) Extraneous variable d) None of the above

32. Which one of the following is independent variable:

    a) Chronological age b) Rate of response
    c) REM rebound d) Reaction time

33. Age is an example of:

    a) Dicotonimous variable b) Discrete variable
    c) Continuous variable d) Intervening variable

34. Sex falls under:
    - **a) Dichotomies variable**
    - b) Polytomies variable
    - c) Continuous variables
    - d) Qualitative variables
35. Example of dichotominous variable is:
    - **a) Male – female**
    - b) Literate – illiterate
    - c) Intelligent – unintelligent
    - d) Introvert – extrovert
36. Which of the following indicates 'Take the odd man out':
    - a) Independent variable
    - b) Active variable
    - c) Stimulus variable
    - **d) Orgasmic variable**
37. An quantitative variable is:
    - a) Skill
    - **b) Age**
    - c) Race
    - d) Religion
38. The word hypothesis indicates:
    - **a) Less then generally held view**
    - b) More than generally held view
    - c) Same as generally held view
    - d) None of the above
39. A hypothesis is a:
    - a) Interrogative statement
    - b) Conditional statement
    - **c) Relational proposition**
    - d) Generally held view
40. For the purpose of rejection, which hypothesis state:
    - **a) Null hypothesis**
    - b) Alternate hypothesis
    - c) Parallel hypothesis
    - d) Both b and c
41. When tested hypothesis found to be true become a:
    - a) Principle
    - **b) Theory**
    - c) Philosophy
    - d) Law
42. The bridge between theory and empirical enquiry is:
    - a) Research principle
    - **b) Research hypothesis**
    - c) Research problem
    - d) Research output
43. Using of one tailed or two tailed depends on the:
    - **a) Alternate hypothesis**
    - b) Null hypothesis
    - c) Composite hypothesis
    - d) Simple hypothesis

44. A researcher moves from hypothesis testing to make prediction, known as:

a) Vertical logic
b) Horizontal logic
c) Inductive logic
d) Deductive logic

45. In research, collection of data falls under:

a) Statistical design
b) Observational design
c) Operational design
d) sampling design

46. Among the following which is the experimental method of data collection:

a) Field study
b) Sample survey
c) Library method
d) Field experiment

47. Survey method mostly used in:

a) Clinical assessment
b) Opinion poll
c) Laboratory experiment
d) Psychoanalysis

48. In a social survey most commonly used method is:

a) Observation
b) Interview
c) Questionnaire
d) Case study

49. Interview method mostly used in:

a) Local studies
b) Public studies
c) Empirical studies
d) Non-scientific studies

50. The style of interview "Hit and Run" is adopted in:

a) Invasion method
b) Immersion method
c) Case method
d) None of the above

51. Alternatives are assigned by values is called:

a) Observation
b) Judgment
c) Measurement
d) Evaluation

52. In social science, measurement is indcates:

a) Objects
b) Properties of objects
c) Phenomena
d) Properties of phenomena

53. The objects character is assign by numbers according to rules is known as:

a) Measurement
b) Evaluation
c) Observation
d) Survey

54. The most lowest/eliminatory level of measurement is:
    a) Nominal b) Ordinal
    c) Interval d) Ratio
55. For measurement of attitude by likert scale falls under:
    a) Nominal b) Ordinal
    c) Interval d) Ratio
56. The lowest level of measurement having an arbitrary zero is:
    a) Nominal level b) Ordinal level
    c) Interval level d) Ratio level
57. Height and weight are examples of:
    a) Nominal level b) Ordinal level
    c) Interval level **d) Ratio level**
58. An example of first psychological test is:
    a) Binet-Simon intelligence scale b) Army Alpha test
    c) Thorndike Handwriting scale d) Woolworths scale
59. Paper pencil test is also called as:
    a) Verbal test b) Non Verbal test
    c) Performance test d) Non-language test
60. An example of rorschah projection tests is:
    a) Association test b) Construction technique
    c) Completion technique d) Ordering technique
61. Thematic Apperception test given basic focus on:
    a) Product b) Process
    c) Input d) Both a and b
62. Thematic apperception test is used to study the:
    a) Leadership behaviour b) Individual creativity
    c) Conflict management **d) Achievement motivation**
63. Projectile test are example of:
    a) Objective test b) Subjective test
    c) Attitude test **d) Both b and c**
64. In role play method which technique is used:
    a) Expressive technique b) Association technique
    c) Constructive technique d) Completion technique

65. In an expressive test, basic emphasis given on:
    a) Product **b) Process**
    c) Input d) Both a and b

66. Who discover the semantic differentiation test:
    a) Fisher **b) Osgood**
    c) Mereno d) William Stephenson

67. The method 'Q Methodology' given by:
    a) Fisher b) Osgood
    c) Mereno **d) William Stephenson**

68. For getting statistical stability and reliability in 'Q Distribution', number of cards require is:
    a) 1-30 b) 31-60
    c) 61-90 d) 91-120

69. The smallest unit of content analysis is:
    a) Theme **b) Words**
    c) Character d) Items

70. In content analysis which computer programme is used:
    a) General Inquirer b) Dream weaver
    c) Mind mapper d) Corel draw

71. The three R's in the SQ3R strategy of comprehension does include:
    a) Read b) Recite
    c) Review **d) All**

72. In explorative studies, observational procedure is more:
    a) Structured b) Unstructured
    c) Partially structured **d) None of the above**

73. Questionnaire involves:
    a) Oral-verbal stimuli **b) Written-verbal stimuli**
    c) Oral-written stimuli d) None of the above

74. A test measures what it is intended to measure is known as:
    a) Reliable **b) Valid**
    c) True d) None of the above

75. If a test score consistent and reproducible, indicated the:
    a) Reliability b) Validity
    c) Credibility d) Sustainability

76. The main property of reliability is:

a) Test score
b) Test as such
c) Norms of test
d) Sample of test

77. If an instrument is valid, indicates that it free from:

a) Variable errors
b) Systematic errors
c) Random errors
d) Non-systemic errors

78. If an instrument is free from variable errors is knows as:

a) Reliability
b) Validity
c) Credibility
d) Sustainability

79. A low error variance indicate that:

a) High validity
b) High reliability
c) Low validity
d) Both a and b

80. In variability alpha coefficient indicates the:

a) Temporal stability
b) internal consistency
c) Both a and b
d) None of these

81. In a perfectly homogenous experimental condition,coefficient of reliability will be:

a) Zero
b) 1
c) 0.5
d) 2

82. The lowest common denominator of a test is called:

a) Sample
b) Item
c) Root
d) Base

83. When we go with high discriminatory items, then:

a) Reliability decrease
b) Reliability increase
c) Reliability not change
d) None

84. The lowest external validity found in:

a) Experimental design
b) Explorative design
c) Diagnostic design
d) Survey design

85. A test showing high error score then the reliability will be:

a) Constant
b) High
c) Low
d) Zero

86. In an laboratory experiment, internal and external validity are respectively:

a) **High and low** b) High and high

c) Low and low d) Low and high

87. The type of validity attached to an instrument is known as:

a) Face validity b) Content validity

c) Criteria related validity d) Construct validity

88. The type of validity that assume most significance when some set standards are compared with the measuring score:

a) Face validity b) Content validity

c) Criteria related validity d) Construct validity

89. On the spot reliability method is:

a) Test re-test method **b) Odd even method**

c) Parallel foams method d) Comparable method

90. The temporal stability and internal consistency dimensions are taken in to consideration in which reliability method:

a) Test retest method b) Odd even method

c) Parallel foam method d) Split half method

91. The statement "All that is reliable is valid, but all those is valid is not reliable" is:

a) Completely false b) Partially true

c) Completely true d) Partially false

92. Jury opinion is a method of establishing:

a) Reliability **b) Validity**

c) Creativity d) Adjustability

93. Attitude comprise of:

a) Belief and emotion b) Emotion and feeling

c) Thought and emotion d) Belief and thought

94. The internal and external validity of field experiment are respectively:

a) High-low **b) Low-high**

c) Low-low d) High-high

95. Among them external validity is highest for:

a) Field experiment b) Field study

c) Sample survey d) Case study

96. The higher validity coefficient in an aptitude test will make a better:
    a) Prediction b) Result
    c) Method d) None of the above
97. In which of the following cases word "Scale" is correctly used:
    a) Achievement scales **b) Attitude scales**
    c) Intelligent scales d) Both a and b
98. Attitude of a people comprise of:
    a) Present experience **b) Past experience**
    c) Future experience d) None of the above
99. By the use of attitude scales, measuring the:
    a) Opinion b) Judgment
    c) Fact d) None of the above
100. The perceived behavior was measured by:
    a) Adorno scale **b) Rating scales**
    c) Both a and b d) None of the above
101. Among them summated rating scale is:
    a) Guttman scale b) Thurstone scale
    c) Adorno scale **d) Likert scale**
102. The basic form of guttman scale is:
    a) Summated rating scale b) Equal appearing interval scale
    c) Cumulative scale d) None of the above
103. Optimum level of difficulty of a test corresponds to:
    a) Maximum item inner-correlation
    b) Maximum item total correlation
    c) Higher reliability of test scores
    d) All of the above
104. If the thurstone and clave scales having lower scale value, indicate the:
    a) Positive attitude b) Negative attitude
    c) Neutral attitude d) Zero attitude
105. Another name of index of discrimination is:
    a) Item validity index b) Item reliability index
    c) Item difficulty index d) Item credibility index

106. Representative sample from a population shows average performance on a given test is known as:

a) Norm	b) Validity
c) Reliability	d) Objectivity

107. Which of the following is an ability test:

a) General intelligence test	b) Social intelligence test
c) Gottshehold test	**d) All**

108. Which test rates a person's potential for learning skills of various occupations:

a) Vocational interest test	b) Interest test
c) Attitude test	**d) Aptitude test**

109. Deriving valid conclusions indicate which branch of knowledge:

a) Logic	b) Paralogic
c) Mnemonics	d) None of the above

110. The word logic originate from root:

a) Greek	b) Latin
c) French	d) Persian

111. Logic is a science of:

a) Value	**b) Foam**
c) Truth	d) Common sense

112. The valid conclusion derives from two or more premises is:

a) Syllogism	b) Paralogic
c) Argument	d) None of the above

113. Prepositions where in the predicate affirms or denies the subjective term is:

a) Hypothetical preposition	**b) Categorical preposition**
c) Disjunctive preposition	d) Relational preposition

114. In a syllogism which appear in both the premises but not in the conclusion:

a) Major term	b) Minor term
c) Middle term	d) All of the above

115. Collection of objects or element is called:

a) Set	b) Union
c) Universe	d) Intersect

116. Manipulated variables are otherwise known as ___________

a) Active b) Concurrent variable

c) Research variable d) Experimental variable

117. Which of the following is also referred as Measured variables?

a) Categorical variable **b) Attribute variable**

c) Concurrent variable d) Manipulated variable

118. Under which principle, every member of a population has an equal chance of being selected ?

a) Principle of replication b) Principle of generalization

c) Principle of randomization d) Local Control

119. Which type of sampling method relates to Quota sampling?

a) Probability b) Choice

c) Non-Probability d) Lottery

120. Concept of 'Law of large numbers' was given by ___________

a) Gosset **b) Jacob Bernoulli**

c) R.A. Fisher d) Charles Spearman

121. Who among the following has divided field studies into two broad types 'exploratory' and ;hypothesis' and 'hypothesis testing' ?

a) Spearman **b) Katz**

c) Likert d) Lindquist

122. 'Rokeach's value survey' is which type of scale?

a) Rank Order b) Nominal

c) Ratio d) Interval

123. In graphic rating scales lines or bars are combined with which phases?

a) summary b) Introductory

c) Descriptive d) Analytic

124. 'Causal Analysis Framework' was developed by which of the following personality?

a) Judea Pearl b) Gabriel Tarde

c) Newcomb d) G J Mendel

125. Which one of the following is another name for Latent Trait Analysis?

a) Response Theory b) Item theory

c) Item Response Theory d) Value addition theory

126. The Rasch model is a psychometric model used for analyzing which type of data?

a) Categorical Data b) Continuous data

c) Omnipresent data d) None of these

127. Which of the following method was introduced by Harold Hotelling in 1936?

a) Canonical-Correlation Analysis b) Factor analysis

c) PCA d) Correlation analysis

128. Common factor analysis, also known in which of the following names?

a) Principle Factor Analysis b) Principal Axis Factoring

c) Both a & b d) None of these

129. Which of the following is not a measure to correct confounding effect.

a) Reliability testing b) Randomization

c) Elimination **d) Addition**

130. The Kelvin temperature scale is an example of which type of measurement?

a) Nominal b) Ordinal

c) Interval **d) Ratio**

131. Under MAXMINCON principle which variance should be minimized?

a) Systematic **b) Error**

c) Extraneous d) None of these

132. Which of the following is true in Random Observation study?

a) No Variable is conceptualised

b) Hypothesis is not formulated

c) No sampling procedure is followed

d) All

133. In which study, as many relevant variables as possible are brought in plan of analysis?

a) Diagnostic study b) Descriptive study

c) Observation study d) Explorative study

134. Which one of the following can be considered as an extension of KR-20?

a) Cronbach's alpha b) Test-retest

c) Spearman Brown d) Tukey's HSD

135. Cronbach's alpha is used to measure ___________

a) Internal consistency
b) Face validity
c) Content validity
d) None of these

136. Bayes' Rule applied for which of the following aspect?

a) Situational probability
b) Systematic probability
c) Conditional probability
d) None of these

137. The term 'percentile' was introduced by whom?

a) F Galton
b) Roger Bacon
c) J.W. Tukey
d) R.A. Fisher

138. In which of the following, reporting format is formally prescribed?

a) Doctoral level thesis
b) Conference of researchers
c) Workshops and seminars
d) Symposia

139. In doing action research what is the usual sequence of steps?

a) Reflect, observe, plant, act
b) Plan, act, observe, reflect
c) Plan, reflect, observe, act
d) Act, observe, plan, reflect

140. Who said that the schedule is nothing more than a list of questions which are answered in order to test the hypothesis?

a) Goode and Hatt
b) Thomas Carson McCormick
c) C.M. Moser
d) Emory S. Bogardus

141. Scale value of a statement in a Thurstone's Equal Appearing Interval Scaling technique is 2.44, then, which of the following is true?

a) Median is less than 2.44
b) Median is more than 2.44
c) Median is equal to 2.44
d) None of the above

142. Define a construct which other construct is known as_______

a) Operational definition
b) Constitutive definition
c) Both a & b
d) Structural

143. Number of pairs in a Paired Comparison test with 5 stimuli will be

a) 15
b) 10
c) 12
d) None of these

144. The proportion of errors in a Scalogram analysis is 0.25. Then, phi coefficient of analysis is

a) 0.50
b) 0.70
c) 0.75
d) None of these

145. During development of a knowledge test, 46 and 60 respondents gave the correct answer of a statement. Then, difficulty index is

a) 45
b) 76.67
c) 15
d) None of these

5

Communication Process & Transfer of Technology (Theory)

Communication word comes from Latin **'Communis'** - Common

Communication: The communication is the process of human interaction in which one person communicates a message/idea to another person.

Diffusion: It is the process by which an innovation/new idea/new technique spreads.

According to Rogers (1983) 'Communication is the process by which the message is transmitted from the source to the receiver'.

Function of communication

1. Informative
2. Instructive/Command
3. Influence/Persuasive
4. Integrative

Communicator: Person who starts the process of communication OR he is the sender of message

Message: It is the information the communicator wishes his audience to receive, understand and act upon

Encoding: It is a process of interpreting the message

Decoding: It is a process of receiving the message

Channel: It is anything which connects the communicator with the group of people OR audience

Treatment: It is the way of handling the message through channel

Audience response: It is the response given by the audience to the message of the communicator

Empathy: The process or ability to project ourselves into others people's personalities/situations.

Credibility: Trustworthiness and competence.

Feedback: Feedback is your audience's response; it enables you to evaluate the effectiveness of communication.

Homophily and Heterophily: Homophily is the degree to which pair of individuals who communicate are similar in certain characteristic such as beliefs, social status etc. Heterophily is opposite to certain attributes.

Communication gap: Gap between what was communicated by sender and what has actually received by the receiver.

Communication fidelity: Faithful performance of communication process by all its elements i.e. Communicator, message, channel and receiver.

Hypodermic needle theory Or Magic Bullet Theory: The "hypodermic needle theory" implied mass media had a direct, immediate and powerful effect on its audiences.

Interpersonal and Intrapersonal communication: In an interpersonal communication takes place both the sender and receiver are present. It means face to face communication. While a communication takes place within a person is called intra personal communication i.e. only one person involved in it. The various situations in which intra personal communication takes place such as thinking while making a decision. In intra-personal communication, the source of the communication is the person himself.

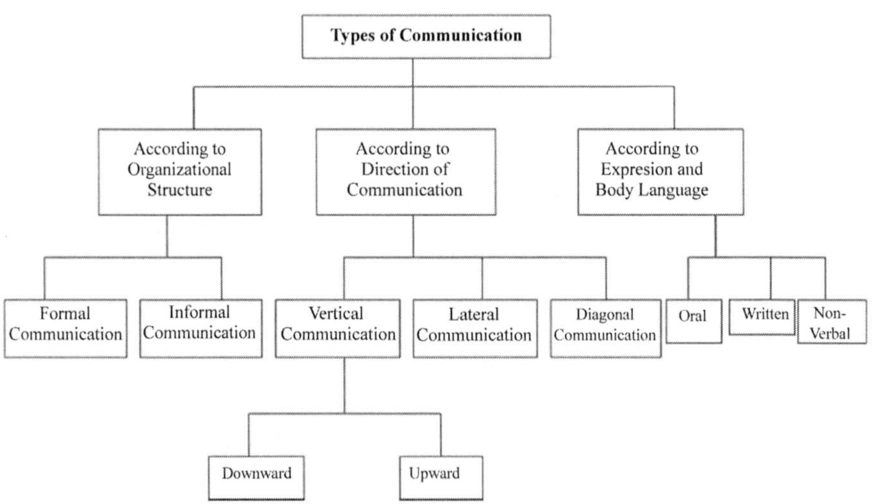

Organizational Communication: It is an evolutionary, culturally dependent process of sharing information and creating relationships in environments designed for manageable, goal oriented behavior in an organization.

Types of Organizational Communication

1. Cross communication – Same level
2. Downward communication – Superior to subordinate
3. Upward communication – Subordinate to superior

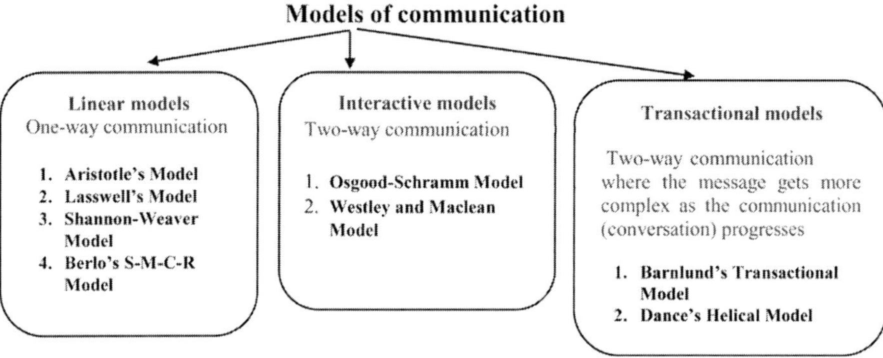

1. **Aristotle's model:** According to Aristotle, communication has three elements

a. Speaker – the person who speaks

b. Speech – the speech that the individual produces

c. Audience – the person who listens

2. **Shannon- Weaver's model, 1949 (Mathematical Theory of Communication)**

a. Source

b. Transmitter

c. Signal

d. Receiver

e. Destination

3. **Berlo's model, 1960**

a. Source

b. Encoder

c. Message

d. Channel

e. Decoder

f. Receiver

Code: It is a system of signals for communication. Encode means to put the message into code.

Channel: The medium through which the signals move,

Decoder: It converts the message in the code into ordinary language which may be easily understood.

4. **Schramm's model, 1961**

a. Source
b. Encoder
c. Signal
d. Decoder
e. Destination

This model of communication is preferably relevant for the mass media.

5. **Leagan's model, 1963**

a. Communicator
b. Message
c. Channel
d. Treatment
e. Audience
f. Response

6. **Rogers and shoemaker's model, 1971** (S-M-C-R-E model), the components of which are :

a. Source
b. Message
c. Channel
d. Receiver
e. Effects

7. **Westley and Maclean's model**

a. Sender
b. Encoder
c. Channel
d. Decoder
e. Receiver

This model can be seen two contexts, interpersonal and mass communication. And the difference between interpersonal and mass communication is the feedback. In interpersonal, the feedback is direct and fast while in the mass, the feedback is indirect and slow.

Theories of communication: Theories of communication classified into three groups namely.

a) Theories of interpersonal communications

1. Kelman paradigm,
2. Bauer's scheme,
3. Festinger's scheme,
4. Attribution theory,
5. Game theory,
6. Social comparison theory,
7. New comb's theory,
8. Stimulus response theory,
9. Linguistic theory,
10. Psycholinguistic theory,
11. Socio-linguistic theory.

b) Theories of mass communication (message flow models):

1. Hypodermic needle model
2. One step flow model
3. Two step flow model
4. Multistep flow model
5. Theory of mass media effects
6. Theory of individual difference
7. Psychodynamic model of persuasion process
8. Theory of social categories
9. Theory of social relationship
10. Free press theory

c) Theories of communication distortion

1. Theory of communication distortion
2. Theory of communication distortion in transit.

Phases of Technology Transfer

1. Technology Innovation
2. Technology Confirmation
3. Targeting Technology consumers
4. Technology Marketing/Extension
5. Technology Application
6. Technology Evaluation

Adoption: A diffusion of innovation with in a social system takes place through its adoption by individual or groups. Adoption is a decision to make full use of an innovation as the best course of action available.

Adoption process: The "adoption process" is a decision-making process goes through a number of mental stages before making a final decision to adopt an innovation.

Diffusion: It is the process by which an innovation is communicated through certain channels over time among the members of a social system. In diffusion, message is new.

Diffusion process: According to Rogers as "the spread of a new idea from its source of invention or creation to its ultimate use of adopters".

Reinvention: It is the degree to which a technology is changed or modified by a user in the process of its adoption and implementation. Re-invention is beneficial to adopters.

Stages of Adoption

a) Awareness Stage - Change agent and mass media
b) Interest Stage - Formal sources (Extension Agency)
c) Evaluation Stage - Fellow farmers and neighbors
d) Trial Stage - Expert and experienced farmers
e) Adoption Stage - Self-experience gained at the trial stage

Rejection: Decision not to adopt an innovation. This may be of two types:

Active rejection: When a farmer rejects after adopting the innovation including even its trial is called active rejection

Passive rejection: Simply non- adoption is called passive rejection.

Dissonance: Human behaviour change is motivated in part by a state of internal disequilibrium or dissonance, an uncomfortable state of mind that the individual seeks to reduce or eliminate.

Discontinuance: Decision to reject as innovation after having previously adopted it. Discontinuance having three forms.

i) Replacement discontinuance is a decision to reject an idea in order to adopt a better idea that supersedes. Eg. Hybrid over variety.

ii) Disenchantment discontinuance is a decision to reject an idea as a result of dissatisfaction with the performance. Eg. Fodder varieties generally deteriorate after number of years.

iii) Forced Discontinuance farmers are forced to discontinue the existing practices because of Government Policies. Eg. The Government has banned the use of chemicals like D.D.T. and B.H.C.

Innovation: "An innovation is an idea/practices perceived as new by an individual.

Innovativeness: It is the degree to which an individual is earlier in adoption when compared with the other members of society.

Innovation Decision Process: According to Rogers 'The Innovation - Decision process is the process through an individual (or other decision making unit) passes from first knowledge of an innovation, to forming an attitude toward the innovation, to a decision to adopt or reject, to implementation of the new idea, and to confirmation of this decision to adopt or reject and to confirmation of this decision'.

The innovation – decision period: The length of time required to pass through the innovation – decision process. The time is in days, months, or years. This period is thus a gestation period in which a new idea is fermenting in the individual's mind. Earlier adopters have a shorter innovation – decision period than later adopter.

Over adoption: Some farmers continue to adopt an innovation, rather vigorously, when experts (Scientists or extension officers) feel that it should be stop.

Preventive Innovations: According to Rogers, 1995 a preventive innovation is an a idea that an individual adopts at one point in time in order to lower the probability that some future unwanted event will occur. Ex: preparing for natural disaster. The benefits of preventive innovations are clearly visible after a lapse of time. A preventive innovation has particularly slow rate of adoption because of individuals have difficulties in perceiving its relative advantage.

Innovators: Innovators are venturesome and first to adopt a new idea, much ahead of other members in the community. They are very few numbers in a social system.

Laggards: Laggards are the last to adopt an innovation. They are traditional. By the time laggards finally adopt an innovation, it may have already been superseded by a most recent idea which the innovators are already using.

Self Image: It is the way we see ourselves.

Legitimization: Certain people or groups that seem to have the right authority and privilege to pass on things to make them justifiable ideas. Ex. Village Officers, Panchayat, President etc.

Grapevine Communication: Informal channel of communication is known as Grapevine.

Social System: "A social system is defined as a population of individuals who are functionally differentiated and engaged in collective problem solving behaviour."

Attributes of Innovation affecting Adoption: Rogers and Shoemaker (1971)

1. Relative Advantages: +VE
2. Compatibility: +VE
3. Complexity: –VE
4. Trialability: +VE
5. Observability: +VE

ADOPTERS CATEGORIES

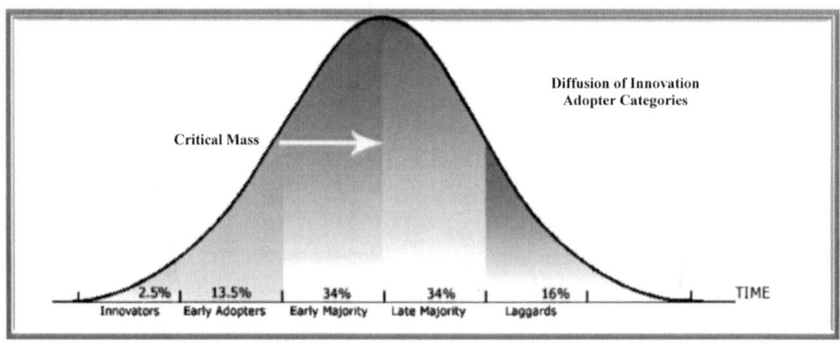

Category	**Innovators**	**Early Adopters**	**Late adopters**	**Late majority**	**Laggards**
Percentage	2.5 %	13.5 %	34 %	34 %	16 %
Characteristics	Venturesome Interested in new ideas	Respectable Greatest degree of opinion leadership	Deliberate Adopt new innovations just before the average member of a social system	Skeptical Adopt new ideas just after the average member of a social system	Traditional Suspicious of innovations Last to adopt an innovation

Agri-Clinics: Agri-Clinics are envisaged to provide expert advice and services to farmers on various technologies including soil health, cropping practices, plant protection, crop insurance, post harvest technology and clinical services for animals, feed and fodder management, prices of various crops in the market etc. which would enhance productivity of crops/animals and ensure increased income to farmers.

Agri-Business: Centres are commercial units of agri-ventures established by trained agriculture professionals. Such ventures may include maintenance and custom hiring of farm equipment, sale of inputs and other services in agriculture and allied areas, including post harvest management and market linkages for income generation and entrepreneurship development.

Agricultural Technology Information Centre (ATIC): The Agricultural Technology Information Centre (ATIC) was started during the year 1999 under the National Agriculture Technology Project (NATP). The Centre provides information on agriculture technologies in addition to providing other inputs like seeds, plant materials, etc. including advisory services through single window system. The Subject Matter Specialists provide technical support to the farmers and others who visit ATICATIC also serves as kissan call centre (Level II) to provide advisory services to the farmers over telephone.

NITI Aayog: National Institution for Transforming India, 1 January 2015.

Farmer FIRST (Farm, Innovations, Resources, Science and Technology) October, 2016.

Student READY (Rural Entrepreneurship Awareness Development Yojana), 25TH July 2015.

Multiple Choice Questions

1. The word 'Communis' is:

   a) Information flow   b) Talk

   c) **Common**   d) None of the above

2. The word communication have its roots:

   a) **Latin**   b) Roman

   c) Greek   d) Persian

3. Communication levels are:

   a) Conventional   b) Exploratory

   c) Participative   d) **All of the above**

4. The important dimensions of message include:
   a) **Code, content, treatment** b) Source, self, receiver
   c) Encode, decode, feedback d) Simplicity and clarity

5. The way of handling a message before placing in the channel known as:
   a) Decoding b) **Treatment**
   c) Distortion d) Feedback

6. Acceptance and rejection are the kinds of:
   a) Failure of channel b) **Feedback**
   c) Treatment of message d) Empathetic feelings

7. In a communication act, degree to which individuals involved differ in certain characteristics:
   a) Homophily b) Credibility
   c) Empathy d) **Heterophily**

8. Example of communication barriers are:
   a) Social b) Semantic
   c) Technical d) **All of the above**

9. Homogeneity is the communication barrier pertaining to:
   a) Source b) Message
   c) Channel d) **Receiver**

10. Communication barrier related to communicator is:
   a) **Provide too much information** b) Homogeneity
   c) Both d) None

11. In mass communication main problem is:
   a) Costly b) **No local dialect**
   c) Same information d) Cumbersome

12. Cybernetics is:
   a) Decision making b) Organizations
   c) Ecology d) **Communication**

13. Indian Institute of Mass Communication located in:
   a) **Jodhpur** b) Chandigarh
   c) Mumbai d) New Delhi

14. Communication breakdown occurs mostly in:
    a) Ulterior transaction
    b) **Cross transaction**
    c) Straight transaction
    d) Vertical transaction
15. Due to which large number of communication models are exist:
    a) **Contextual variation**
    b) Behavioral variation
    c) Sequential variation
    d) Geographical variation
16. In which model of communication 'Communicator' is the characteristic element:
    a) **Leagans**
    b) Lazwell
    c) Berlo
    d) Rogers
17. The book entitled "The theories of mass communication" is written by:
    a) Gerbner
    b) **Defleur**
    c) Mequail
    d) Windehl
18. The smallest meaningful unit of speech is:
    a) Heuristic
    b) Productive
    c) **Morphemes**
    d) Phonemes
19. Who introduced the concept of feedback?
    a) **D Berlo**
    b) P Leagans
    c) Hovland
    d) Westley and Mclean
20. The Berlo's model of communication contains:
    a) 4 elements
    b) 5 elements
    c) **6 elements**
    d) 7 elements
21. Aristotle's model of communication is contains:
    a) Speaker
    b) Speech
    c) Listener
    d) **All of the above**
22. The S-M-C-R-E model of communication it given by:
    a) Berlo
    b) Laswell
    c) **Rogers**
    d) Leagans
23. In which communication model 'Destination' is an important element:
    a) Leagans
    b) Berlo
    c) **Schramm**
    d) Laswell

24. In which communication model 'Audience response' is an important elementl:

a) **Leagans** b) Berlo

c) Schramm d) Laswell

25. Role taking theory was given by:

a) Solaman Asch b) **George Mead**

c) Morgan and King d) Chitambar

26. Who says that "Each person in the communication process is both an encoder and a decoder":

a) **Schramm** b) Leagans

c) Berlo d) Auxinn G H

27. Communication IQ was developed by:

a) **Bhanja** b) Paul Vegans

c) Hovland d) None of the above

28. Berlo given the types of skills:

a) **2** b) 3

c) 5 d) 4

29. Hypodermic needle theory employed for:

a) Individuals b) Group

c) **Mass** d) None

30. The first basic persuasive communication model:

a) **D Berlo** b) Wendell

c) Wilber Schramm c) Richard Dance

31. Mathematical theory of communication was given by:

a) Aristotle b) **Shannon and Weaver**

c) Lasswell d) Berlo

32. In which communication model noise is an important element:

a) Aristotle b) **Shannon and Weaver**

c) Lasswell d) Berlo

33. The term key communicator was first used by:

a) E M Rogers b) **Lazerfield**

c) D Berlo d) Stevens

34. How much percentage of communication takes place through body language:

a) **40** b) 30

c) 20 d) 10

35. Full form of LCD:

a) Lazier Crystal Digital b) Lazier Cystal Display

c) **Liquid Crystal Display** d) Liquid Crystal Digital

36. The electronic chalkboard also known as:

a) LCD projector b) **OHP projector**

c) VHS projector d) Beta chrom projector

37. Diascope when changed to epidiascope will change the projection from:

a) Reflected to direct b) **Direct to reflected**

c) Dircct to indirect d) Indirect to direct

38. If a hall length 25m., then the ideal letter size for a poster:

a) 3.5 inch b) **2.7 inch**

c) 2.1 inch d) 1.4 inch

39. Commonly used film projector in extension:

a) 8 mm b) **16 mm**

c) 32 mm d) 70 mm

40. The slide – cum-film projector work on the principle of:

a) **Direct projection** b) Indirect projection

c) Reflected projection d) Diffused projection

41. Maximum distance up to which one inch object can be seen clearly:

a) 2.43 m b) 4.8 m

c) 7.31 m d) **10.97 m**

42. The object position in slide projections should be:

a) **Inverted** b) Vertical

c) Horizontal d) Diagonal

43. Which instruments used the Reflect projection?

a) OHP b) Slide projector

c) Lantern projector d) **Epidiascope**

44. Amplifier is an example of:
    a) **Electronic medium** b) Print medium
    c) Written medium d) None of these
45. In a chart most appropriate letter size for titles is:
    a) 1.5" to 2" b) **2.5"**
    c) 1" to 2" d) 2"
46. The most suitable seating arrangements for the motion pictures is:
    a) **2 and 6 formula** b) 4 and 6 formula
    c) 2 and 4 formula d) 1 and 5 formula
47. For a group of 30 to 50 persons, size of flash cards should be:
    a) 10 x 12 inches b) **15 x 20 inches**
    c) 20 x 25 inches d) 20 x 30 inches
48. The 'COIK' fallacy related with:
    a) Communication channels b) **Audio Visual aids**
    c) Communicator d) Receiver
49. All capital words are not read easily due to:
    a) **Lack of symmetry** b) Uneven loops and strokes
    c) Psychological problem d) None of the above
50. The ratio between 'F' number and shutter speed of a camera is:
    a) 1:2 b) 1:3
    c) **2:1** d) 3:1
51. An example of primary colour is:
    a) Pink b) **Red**
    c) Violet d) Orange
52. Which colour have highest value:
    a) Red b) Green
    c) **White** d) Yellow
53. For extension work, primary colours are:
    a) **Red, Blue, Yellow** b) Red, Blue, Green
    c) Red, Blue, Purple d) Red, Yellow, Green
54. For clear visualization, distance between two lines of materials is:
    a) 1/5th of the letter height b) 2/5th of the letter height
    c) **2/3rd of the letter height** d) 3/5th of the letter height

55. The leaflet is a:
    a) **Single sheet of paper** b) Four sheet of paper
    c) Six sheet of paper d) Eight sheet of paper
56. Usually flash card used in numbers of:
    a) **10-12** b) 12-14
    c) 14-16 d) 16-18
57. Which teaching aid used for creating suspense, sequence and storytelling effects:
    a) Flash card b) Flannel graph
    c) Flip chart d) **All of the above**
58. The ABC in journalism indicates:
    a) Accuracy, Brevity, Credibility b) **Accuracy, Brevity, Clarity**
    c) Accountable, Brief, Clear d) Active, Brief, Clear
59. The news story headline is generally given in:
    a) **Present tense** b) Past tense
    c) Future tense d) No tense
60. Small drawing at the end of a story is known as:
    a) **Tailpiece** b) Legend
    c) Masthead d) Quote
61. In India, first farm magazine is:
    a) **Kethi** b) Phool Phul
    c) Krishi Chayaika d) Yojana
62. Press trust of India was established in:
    a) **1949** b) 1951
    c) 1960 d) 1972
63. The national press day is:
    a) 18th June b) 06th May
    c) 12th October d) **16th November**
64. First newspaper of India was:
    a) **Bengal Gazette** b) Rural India
    c) Times of India d) Bengal Outlook
65. The programme "Krishi Dharshan" is telecasted from:
    a) Mumbai b) Kolkata
    c) **New Delhi** d) Banglore

66. Publication of 'Rural India' is from:
    a) New Delhi   b) **Poona**
    c) Meeret   d) Hyderabad
67. Publication of Bengal Gazette is from:
    a) **Calcutta**   b) Bombay
    c) Madrass   d) Delhi
68. Prasar Bharti started on:
    a) 15th October 1995   b) 18th January 1996
    c) 20th December 1996   d) **23rd January 1997**
69. A farm publication used to provide precise and reliable scientific information in simple language about a practice is known as:
    a) **Leaflet**   b) Literature
    c) News paper   d) Circular letter
70. The first farm periodical is:
    a) Kheti   b) **Krishi-sudhar**
    c) Bhumi-suchar   d) Kurukshethra
71. Broadcasting through radio started in India on:
    a) June 17, 1923   **b) July 23, 1927**
    c) July 27, 1925   d) July 17, 1928
72. Broadcasting through radio started in India from:
    a) Delhi   b) Banglore
    c) Calcutta   d) **Mumbai**
73. The Akashwani came in to existence in the year:
    a) 1947   b) 1950
    c) **1957**   d) 1960
74. Radio is come under:
    a) **Hot media**   b) Cold media
    c) Optical Media   d) Fiber media
75. In a radio talk, rate of delivery of words should be:
    a) 80-100 words/minute   b) **120-140 words/minute**
    c) 160-180 words/minute   d) 200-220 words/minute

76. In India, telecast on television was started on:
   a) 15th August 1959 b) **15th September 1959**
   c) 15th October 1959 d) 15th November 1959
77. DD-Gyandarshan, was started in India on:
   a) 15th August 1999 b) 26th January 1999
   c) **26th January 2000** d) 15th August 2000
78. The organization, interpretation, reformulation and translation of research knowledge into technology happened in:
   a) **Technology generation** b) Technology production
   c) Technology integration d) Technology testing
79. During process of technology transfer, function of extension starts from:
   a) **Technology testing** b) Technology production
   c) Technology integration d) Technology dissemination
80. The linkage between research and extension essential overlap on:
   a) Technology testing b) Technology production
   c) Technology integration d) **All of the above**
81. In the farmer first model technology flow is basically form:
   a) Agent to farmer b) Researcher to agent
   c) **Farmer to farmer** d) Researcher to farmer
82. Technology evaluation indicates:
   a) **Value judgment** b) Valere
   c) Positive and negative d) None of the above
83. Idea by which transfer of technology is related:
   a) Local Idea b) **Progressive idea**
   c) Foreign idea d) None of the above
84. In a social system, spread/dissemination of technology by:
   a) **Diffusion** b) Teaching
   c) Learning d) Adoption
85. The Rogers model of diffusion assumes that:
   a) **The homogeneity of population**
   b) The progressiveness of farmers
   c) Availability of infrastructural facilities
   d) The differential farming situation

86. Diffusion of an new technology is decided by:
    a) **Society**
    b) Individual
    c) Both
    d) None

87. Diffusion is a subset of:
    a) **Communication**
    b) Teaching
    c) Learning
    d) Motivation

88. Full form of CONA is:
    a) Career Oriented Need Assessment
    b) **Community Oriented Need Assessment**
    c) Creative Organizational Need Assessment
    d) Career Oriented Need Analysis

89. The new idea is:
    a) Information
    b) **Innovation**
    c) Perception
    d) Invention

90. In case of innovation, symbolic adoption takes place with:
    a) **Idea component**
    b) Object component
    c) Product component
    d) Relative component

91. The character of an innovation which is important form extension point of view:
    a) Intrinsic
    b) **As people see them**
    c) Extraordinary
    d) Typological

91. An innovation can be originated from:
    a) Farmer research
    b) Extension programme
    c) Research trails
    d) **All of the above**

92. The newness of an innovation can be seen in terms of:
    a) Knowledge
    b) Decision to adopt
    c) Persuasion
    d) **All of the above**

93. Among them which perceived attribute of an innovation reduces the risk component:
    a) Complexity
    b) Compatibility
    c) **Trialability**
    d) Observability

94. The low relative advantage with an innovation cause:

a) Slow rate of adoption
b) High rate of discontinuance
c) Low return on investment
d) **All of the above**

95. Compatibility is the degree to which an innovation is perceived consistent with:

a) Peoples needs for innovation
b) Previously introduced needs
c) Socio-cultural values and beliefs
d) **All of the above**

96. The less adoption rate in 'Preventive innovations' mainly due to:

a) Low relative advantage
b) High complexity
c) Low trialability
d) **Low observability**

97. Among them which perceived attribute of an innovation shows a negative correlation with adoption:

a) **Complexability**
b) Compatibility
c) Trialability
d) Observability

98. In the process of time is a very important consideration:

a) Communication
b) Adoption
c) **Diffusion**
d) Innovation

99. In a social system, adoption of innovations by all individuals occurs in:

a) At the same time
b) At equal time intervals
c) **An ordered time sequence**
d) Time hardly matters

100. Innovation decision period is also called as:

a) **Gestation period**
b) Production period
c) Mental period
d) None of these

101. Role of 'Field of Experience' was described in whose model of communication?

a) Aristotle
b) Schramm
c) Shannon and Weaver
d) Berlo

102. Hovland designed a persuasion model of communication by following whose idea.

a) Aristotle
b) New Comb
c) Leagans
d) Berlo

103. Which distortion gives extra unwanted information?

a) Fog
b) Mirage
c) Systematic
d) None of these

104. Relationship between size of a group and empathetic accuracy is ______

a) **Inverse** b) Positive

c) Both a & b d) Not related

105. Famous theory of instinct of imitation was given by __________

a) McDonald b) **Gabriel de Tarde**

c) McCormick d) Hovland

106. Baa of sheep, moo of cow, quack of duck are the examples of

a) Semiotics b) **Onomatopoeia**

c) Semantopeia d) Myopia

107. Lexicology is the part of linguistics which studies ____________

a) **Words** b) Letters

c) Sentence d) Verbs

108. The term "biosemiotic" was first used by which of the following?

a) **Friedrich S. Rothschild** b) Rutherford

c) Maslow d) Robert Redfield

109. The study of animal communication is also known as _________

a) Pragmatic b) Biosemiotic

c) Semeiotics d) **Zoosemiotics**

110. No time for students to read printouts, reports, projections and forecast is known as ___________

a) **Information indigestion** b) Information overlooked

c) Information overload d) Worn out behavior

111. Walter Lippmann's barriers to effective communication include which of the following aspect (s)?

a) Artificial censorship b) Gatekeepers in the media

c) Shrinking news holes d) **All the above**

112. The space around the speaker is called ___________

a) **Proximics** b) Kinesics

c) Conciseness d) Deciphering

113. Which of the following is incorrectly matched with respect to massage distortion?

a) Fog distortion-Information is marked

b) Systemic distortion-No information is lost

c) Mirage distortion-Unwanted information comes in

d) **Stretch distortion – Full information is lost**

114. 'Mind, self and society' – this book has been written by –

a) E.L. Thorndike b) O.H. Benson

c) George H. Mead d) Paul Hoffman

115. Ornament or possession that communicates information about a person, is known as

a) Museum **b) Artifact**

c) Collection d) Vault

116. A set of rules which is used to translate a message from one form to another is known as

a) Code b) Encryption

c) Description d) Solving mystery

117. Which branch of philosophy investigates the interpretation of texts?

a) Hepatics b) Citadel

c) Phonemics **d) Hermeneutics**

118. Who developed the concept 'Mcworld'?

a) Adam Amith b) Mc Donald

c) Benjamin Barber d) Richmond Greggory

119. Positive classroom communication leads to

a) Coercion b) Submission

c) Confrontation **d) Persuasion**

120. Classroom communication is the basis of

a) Social Identity b) External inanities

c) Biased passivity d) Group aggression

121. If two standard form categorical propositions with the same subject and predicate are related in such a manner that if one is undetermined the other must be undetermined, what is that relation?

a) Contrary b) Sub contrary

c) Contradictory d) Subaltern

122. In which teaching method learner's participation is made optimal and proactive?

a) Discussion method b) Buzz section method

c) Brainstorming session method d) Project method

123. Encoding skills includes __________

a) Writing
b) Painting
c) Reading
d) Both a & b

124. Who developed Media Dependency Theory?

a) Kartz
b) McCOMBS
c) Ball-Rokeach and DeFluer
d) Shaw

125. SPRING Network works in collaboration with

a) DFID and Nike Foundation
b) USAID and DFAT
c) Both a & b
d) None of these

126. The design that reduces the uncertainty in the cause-effect relationships involved in advising farmers and desired outcome is referred as __________

a) A technology
b) Folkways
c) Norms
d) Idea

127. Heart of diffusion process is

a) Advertisement
b) Selling
c) Imitation
d) Farmer

128. Choices to adopt or reject an innovation are made by consensus among the members of a system are called as __________

a) Optional innovation-decisions
b) Collective innovation-decisions
c) Authoritative
d) Forced innovation- decisions

129. Which one of the following lacks personal touch?

a) Documentary
b) Telephone
c) Advisory letter
d) All of the above

130. The social system's norms serve as ______ on the individual's decisions.

a) Incentives
b) Restraints
c) Either a or b
d) None of these

131. What is a state of dissatisfaction or frustration that occurs when one's desires outweigh one's actualities?

a) Need
b) Wish
c) Motives
d) All the above

132. According to Rogers how many stages are involved in the Innovation – Decision Process?

a) Three | b) Seven
c) Five | d) Nine

133. Which one of the following is not the perceived attributes of an innovation?

a) Relative advantage | b) Compatibility
c) Complexity | **d) Innovativeness**

134. __________ Is a decision to reject as innovation after having previously adopted it.

a) Direct rejection | b) Indirect rejection
c) Discontinuance | d) None of these

135. When a farmer who adopts the practice do not influence the other farmers to do the practice is called as __________

a) Passive adoption | b) Active adoption
c) Over adoption | d) Under adoption

136. Which of the following is related to Late majority?

a) Venturesome | b) Respectable
c) Deliberate | **d) Skeptical**

137. The adoption of an innovation over time follows which type of curve, when plotted over time on frequency basis?

a) Bell-shaped curve | **b) S-shaped curve**
c) U-shaped | d) L- shaped

138. Multiple use of an innovation may be a form of __________

a) Relative advantage | b) Compatibility
c) Complexity | d) Observability

139. Which of the following attributes is having negative relationship with rate of adoption?

a) Compatibility | **b) Complexity**
c) Obsevability | d) Relative Advantage

140. __________ refers to the degree or certainty of receiving expected benefits from the adoption of an innovation.

a) Trialability | b) Observability
c) Predictability | d) Relative Advantage

141. Through which element we can easily differentiate diffusion and communication?
    a) Innovativeness
    b) Message
    c) Channel
    d) Time
142. The first study of diffusion was conducted on hybrid seed corn in 1943 by ___________
    a) Rogers
    b) Ryan and Gross
    c) Gabrial Tarde
    d) Lasswell
143. The average time of adoption by the members of a social system will follow which of the following?
    a) Always vary
    b) Sometime vary
    c) Same for all
    d) None of these
144. Crop insurance is the example of which of the following attributes ?
    a) Preventive innovation
    b) Predictability
    c) Trialability
    d) Relative Advantage
145. Accessing email through mobile phone by using 4G network, instead of using computer or laptop is an example of
    a) Discontinuous innovation
    b) Upgraded innovation
    c) Discrete innovation
    d) Up gradation innovation
146. Who has given the 'Chain linked model' of innovation?
    a) J.W. Hawkins
    b) Stephen J.Kline
    c) E.H. Rudolph
    d) K.E. Sumner
147. For ranking countries on the basis of innovativeness which of the following index is used?
    a) Universal Innovation Index
    b) Bloomberg Innovation Index
    c) UN Innovation Index
    d) None of these
148. According to Global innovation Index (2016), which of the following country rank first in the list?
    a) Denmark
    b) Switzerland
    c) Ireland
    d) USA
149. In which year, Advanced Centers of Training (ACT) in India was launched?
    a) March, 1988
    b) January, 1992
    c) March, 1998
    d) January, 1982

150. Which of the following training method can be considered as modified version of Panel?

a) Colloquy b) Seminar

c) Conference d) Lecture

151. Video-text can be considered of which of the following?

a) Tele-text b) View-data

c) Both a & b d) None of these

152. Who has given the idea of meta-evaluation?

a) Michel Scriven b) Lynton and Pareek

c) Iceberg d) Charles Darwin

153. Which of the following uses the concept of empathy?

a) Buzz session b) Seminar

c) Symposium **d) Role playing**

154. The Discrete Phases Approach to training is also known as __________

a) Traditional b) Modern Model

c) Medieval Model d) Historical Model

155. The process of determining a trainee's a achievement in relation to other trainees is known as __________

a) Performance b) Competition

c) Norm referenced measurement d) Progress

156. When a separate work area is set up for a learner so that the actual work situation does not pressure the learner, that type of training is known as

a) Capsule training b) Work training

c) Vestibule training d) Job training

157. How many number of elements involved in training?

a) 4 b) 5

c) 6 d) 7

158. In KASA __________ Last A stands for

a) Attribute b) Attitude

c) Aspiration d) Action

159. Transactional Analysis was founded by

a) Eric Berne b) Dr. Wilder Penfield

c) Thomas A. Harris d) John Dalton

160. Which of the following is related to skill teaching
   a) Farm and home visit b) Radio talk
   c) Media forum d) All of the above
161. NASSDOC stands for
   a) National Science Documentation Centre
   b) National Social Science Digitization Centre
   c) National Social Science Delivery Centre
   d) National Social Science Documentation Centre
162. The term 'Journ' means
   a) Week **b) Day**
   c) Piece of paper d) Communication
163. Television with its unique potential can communicate with which of the following senses?
   a) All the senses **b) Eyes and Ears**
   c) Ear d) Eyes
164. Advocacy Journalism was introduced in which of the following country?
   a) India **b) USA**
   c) Russia d) Germany
165. Indian Institute of Mass Communication (IIMC) was established in 1965 at New Delhi on the recommendation of __________
   a) Asian Development Bank b) UNESCO
   c) Ford Foundation **d) Both b & c**
166. Yellow Journalism is also denoted by which of the following?
   a) Colored Journalism b) Pale Journalism
   c) Sensationalism d) Blind Journalism
167. The first radio programme was broadcasted in the year 1923 by which of the following?
   a) Radio club of Calcutta b) Radio club of Delhi
   c) Radio club of Bombay d) Radio club of Patiala
168. INSDOC stands for
   a) Indian National Scientific Documentation Centre
   b) Indian Serial Documentation Centre
   c) Indian National Scientific Digitalization Centre
   d) Indian Scientific Documentation Centre

169. In the British system of proof reading 'np' is used for

a) New Paragraph b) Next Page
c) New Page d) Next Paragraph

170. The person who marks the corrections on the proof is called

a) Marks man b) Proofer
c) Proof reader d) Corrector

171. In day-to-day conversation all of us use which of the following voice?

a) Active Voice b) Passive voice
c) Inert voice d) Indicative voice

172. Coloured telecast in India started from which date?

a) 2nd October, 1980 **b) 15th August, 1982**
c) 26th January, 1985 d) 1st May, 1975

173. The first farm magazine in agriculture which was published in Paris in the year 1763, was known as

a) Agri journale b) Agri De Review
c) Agriculture De Paris **d) Agriculture De Franche**

174. "Basundhara" and "Krishik Jagatt" fam magazines were published in which language?

a) Bengali b) Marathi
c) Gujarati d) Oriya

175. The other name of content analysis is

a) Concept analysis b) Module analysis
c) Structural analysis **d) Textual analysis**

176. Who developed the first reading ease formula?

a) Bertha A. Lively b) Sidney L. Pressey
c) Both a & b d) None of these

177. Which one of the following index measures the readability of English writing?

a) GE Index **b) Gunning fog index**
c) Mirage index d) Countability index

178. "The Flesch-Kincaid" Formula was first used by whom?

a) US army b) Mc Graw Hill Company
c) British army d) Australian Army

179. Which one of the following was the first printed periodical in the world?

a) The Hindu
b) Bengal gazette
c) Mercurius Gallobelgicus
d) Readers Digest

180. Agricultural Research Information Centre was established by ICAR in which year?

a) 1967
b) 1975
c) 1977
d) 1957

181. In the 'KISS' principle of writing K stands for –

a) Keep
b) Know
c) Knowledge
d) Kind

182. The Prasar Bharati (Broadcasting Corporation of India) Act, was enacted in which year?

a) 1995
b) 2001
c) 1990
d) 1992

183. Asia-Pacific Institute For Broadcasting Development (AIBD) is headquartered at

a) Kuala Lumpur
b) Jakarta
c) Beijing
d) Colombo

184. Kisanvani project was launched in which year?

a) December, 2004
b) February, 2004
c) December, 2005
d) November, 2001

185. ISSN stands for __________

a) International Standard Serial Number
b) Indian Standard Serial Number
c) Indian Standard and Safty Number
d) Indian Standard System Number

186. Modern journalism deals with five departments of mass communication which includes

a) Radio, Newspaper, Internet, Cinema and traditional arts
b) Newspaper & periodicals, Radio, Television, Films and Advertising
c) Music, Film, Internet, Mobile and Photography
d) Photography, Classical dance, Film, Internet and Publicity

187. Principles/canons of journalism are __________
   a) Sincerity and accuracy
   b) Impartiality
   c) Fair play and decency
   d) All of the above

188. Who started first Vernacular periodical-Digdarshan-Bengali in 1818?
   a) Ravindra Nath Tagore
   b) S.C. Bose
   c) J. C.Raman
   d) Raja Ram Mohan Roy

189. 1st Hindi paper Udunt Martund in Calcutta was printed in which year?
   a) 1822
   b) 1829
   c) 1825
   d) 1826

190. Central Bureau of Information (Now PIB) was started in __________
   a) 1921
   b) 1920
   c) 1919
   d) 1925

191. Copyright Act was enacted in __________
   a) 1960
   b) 1947
   c) 1957
   d) 1967

192. Press Council was set up in __________
   a) 1965
   b) 1956
   c) 1976
   d) 1966

193. Samachar Bharti – a multilingual news agency was set up in __________
   a) 1967
   b) 1965
   c) 1977
   d) 1969

194. The structure of the feature article should contain
   a) Title
   b) Body
   c) Lead
   d) All of the above

195. Circulation of farm magazines in descending order is __________
   a) Hindi> English>Bengali>Telugu
   b) Hindi>Bengali>English>Telugu
   c) Hindi>English>Tamil>Telugu
   d) Hindi>Marathi>Bengali>Telugu

196. Kisan Bharti and Farm Digest are published by __________
   a) G.B Pant University of Agriculture and Technology
   b) Punjab Agricultural University
   c) Odisha University of Agriculture and Technology
   d) Allahabad Agricultural Institute

197. Including authors in an article who did not contribute significantly is known as __________

a) Gift authorship
b) Ghost authorship
c) Loan authorship
d) Give and take policy

198. What point should be considered while choosing the right journal for publication?

a) Is the Journal peer-reviewed
b) Who is this journal's audience
c) What is the journal's Impact Factor
d) All the above

199. First Extension Journal in the world was

a) Journal of rural extension service
b) Extension education journal
c) The journal of cooperative extension service
d) Indian Journal of extension education

200. Which of the following term is used for the space utilized in a newspaper allotted to news and illustrations?

a) News point
b) News hole
c) Black hole
d) Star point

201. What is the term used for the page opposite to editorial page?

a) Crown page
b) Baed page
c) Oped page
d) Page 3

202. The information which is passed to a reporter confidentially is known as

a) Tip
b) Top
c) Sincere
d) None of these

203. When a short and related story is added to the end of a long story, then it is known as

a) Tail end
b) Shirttail
c) Fag end
d) Curtail

204. What is the common term used for extended unwanted silence in radio?

a) Dead air
b) Sleep time
c) Zero noise
d) Silence

205. Which of the following term is used to delete whole or part of news story?

a) Delete **b) Cut**

c) Kill d) Short

206. 'Cloze procedure' which is a newtool for measuring readability has been developed by?

a) Fredric Wilon **b) Wilson Taylor**

c) Richard Cloze d) William Cloze

6

Extension Teaching Methods (Theory)

Extension teaching methods: Methods by using them new information can pass freely between the extension worker and rural communities. Extension education is mainly concerned with the dissemination of useful and practical information relating to rural life and helping the rural people in the practical application of such knowledge to rural situations. Extension teaching methods can be classified on the basis of:

Classification of Extension Teaching Methods (According to Wilson & Gallup 1953)

According to use		
Individual contacts	**Group contacts**	**Mass contacts**
Farm and home visits	Methods demonstration, meetings	Bulletins
Office calls	Leader training meetings	Leaflets
Telephone calls	Lecture meetings	News letters
Personal letters	Conferences and discussion meetings	Circular letters
Result demonstration	Meetings at result demonstrations	Radio
	Tours	Television
	Schools	Exhibits
		Posters

According to form			
Written	**Spoken**	**Visual or Objective**	**Spoken and visual**
Bulletins	General special meetings	Result demonstrations	Method demonstration motion
Leaflets	Farm and home visits	Exhibits	Meetings at result demonstrations
News articles	Office calls	Posters	Meetings with motion pictures, charts etc.
Personal letters Circular letters	Radio and Telephone calls	Motion pictures, charts, slides, and other visual aids	Television

Result Demonstration vs Method Demonstration

S. No.	Contents	Result Demonstration	Method Demonstration
1	Purpose	To show locally the worth or value of a recommended practice.	To teach how to do a job involving skill.
2	Conducted by	Farmer under the guidance of extension worker	Extension worker himself or local leader.
3	Benefit of	The demonstrator as well as other farmer	Persons present at the demonstration
4	Comparison	Essential	Not essential
5	Time required	Substantial period	Relatively very little
6	Cost	Costly	Relatively cheap

Some Definition

Farmer's call: Call made by a farmer or homemaker at the working place of the extension agent for getting information.

Adoptive/minikit trial: It is a method of determining the suitability of a new practice in farmer's situation.

Farm clinic: It is a facility developed for diagnosis and treatment of farm problems and provides specialist advice to individual farmers.

Field day/ Farmer's day: Methods of motivating the people to adopt a new practice by showing what has actually been achieved by applying the practice under field condition.

Campaign: It is an intense educational activity for motivating and mobilizing a community to action, to solve a problem or satisfy a need urgently felt by it.

Exhibition: An Exhibition is a systematic display of models, specimens, charts photo etc. in a sequence around a theme to create awareness and interest in the community.

Radio rural forum: A combination of mass media and inter personal communication.

Farm school on the air: It is a method of providing systematic education on farming to the farmers through the process of distance learning.

Audio visual aids: Instructional device which are used to communicate message more effectively through sound and visuals. Mostly used in extension teaching.

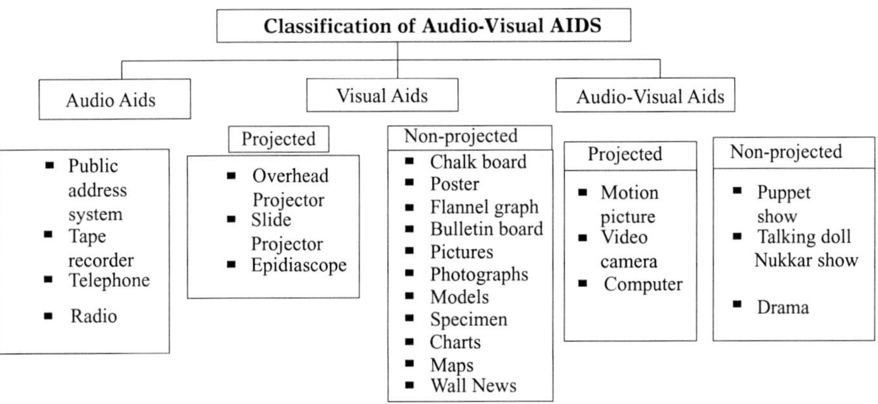

Diagram: It is line drawing of an object or an idea.

Map: It is an informative diagram of an area

Chart: It contains information in tabular form

Graph: It is a diagrammatic representation of the relationship between variables

Specimen: It is a sample, which represents the whole.

Model: It is a miniature replica of an object.

Diorama: It is a scenic representation of the original, with specimen, model, and painting. The word diorama is derived from the Greek Word which means 'to see through'.

Flash cards: The non-projected visual aid in which series of illustrated cards are flashed as presented in a sequence to explain complete story. Total number of flash cards must be 10 -12 which are easy to remember to farmers. Flash card should be large enough for everyone to see at least 22 x 28 inches.

Flannel graph: In flannel graph, paper containing the message is prepared with sand paper backing and is placed one by one in a sequence along with talk, on a board cover with flannel or hand woven thick cloth (khaddar graph).

Flip chart: Flips carry a series of ideas arranged sequentially. Individually charts are tagged or bound to some support. During teaching, these flips are turned one by one in a sequence.

Slide projector: Based on the principle of direct projection.

Over Head Projector: A projector, which projects the image over the head of an instructor on the screen, is called OHP. It worked on the principle of indirect projection.

Extension Literature

Leaflets: A leaflet is usually a single sheet of printed matter, sometimes folded. It gives accurate or specific information on a particular topic.

Pamphlets: The pamphlet containing 2 to 12 pages.

Circular Letters: Circular letters give an intimate, personal approach to the extension message. It can be very effective if it has a personal touch and drawings, cartoons or pictures. There are two types of circular letters:

1. Announcement
2. Subject Matter

Newsletters: A newsletter deals with current items of interest and normally is printed periodically. Newsletters deal with more than one subject, are printed in large numbers and distributed in many ways.

Folders: A folder is a single piece of paper folded once or twice. When opened, material is presented in sequence. Make sure this sequence appears in the finished folded side.

Bulletins: A Bulletin conveys large amount of information. Its primary objective is to give information which the reader can apply to his own local situation. It is a booklet running into more than 20 pages. Bulletins are of two kinds:

1. Technical bulletin – designed primarily to present scientific material to those working in specific fields.
2. Popular bulletin – to present material to people in the field of extension.

Newspaper: It is a bunch of loose printed papers properly folded, which contains news, views, advertisements etc and is offered for sale at regular intervals, particularly daily or weekly.

Banners: Banners are traditional mass media information visuals in India.

Community radio: Community radio is characterized by access, public participation in production and decision-making, and by a non-profit listener-friendly economics. It is a broadcasting organization established to provide communication support for the social, economic and cultural development of a community within a geographical location and owned and operated by the community on a non-profit basis.

Stakeholders: Stakeholders are those who has some interest in running the programme, and a 'stake' in the project. They are all partners in the project and share the responsibility of making it successful.

Folk media: Folk Media is the creative dissemination of information through cultural and performance arts. In traditional societies, folk media: drama, skits, poems, stories, riddles, songs and dance have been popularly and successfully used to disseminate messages and even to pass on wisdom of older generations to the youth.

Multiple Choice Questions

1. The size of class room slide is:
   a) 20 mm b) 120 mm
   c) **35 mm** d) 42 mm
2. Under Krishi Vigyan Kendra training to rural people given on:
   a) Agricultural b) Home science
   c) Animal Husbandry d) **All**
3. Projected visual aid is:
   a) **Over head projector** b) Radio
   c) Flash card d) Banner
4. Study tour is an example of:
   a) Individual contact b) Mass contact
   c) **Group contact** d) None
5. During extension teaching the teacher should know the:
   a) Subject matter b) Teaching aids
   c) Place of teaching d) **Attitude of the learner**
6. Who was introduced the bond theory of learning:
   a) C L Hull b) Ivan Pavlov
   c) **E L Thorndike** d) None of these
7. Which of the following considered both spoken and visual extension teaching method:
   a) Posters b) Radio
   c) **Campaign** d) Agricultural clinics
8. Type of flag method is:
   a) **Spoken method** b) Written method
   c) Group method d) Mass method
9. The extension method which reduced the tension:
   a) **Group discussion** b) Method demonstration
   c) Result demonstration d) Campaign

10. Result demonstration is operationalized by:
    a) Researcher b) Extension worker
    c) **Farmer** d) All of the above
11. An example of non projected visual-aid is:
    a) OHP c) Slide projector
    b) Epidiascope d) **Banner**
12. Newspaper is an example of:
    a) Personal contact c) **Mass contact**
    b) Group contact d) None of the above
13. Which of following projector based on reflected projection?
    a) OHP c) Slide projector
    b) Film projector d) **Epidiascope**
14. In opaque projection which of the following equipment is used?
    a) OHP c) Slide Projector
    b) Epidiascope d) Film projector
15. For rural development in India, basic operational unit is:
    a) District c) Village
    b) Block d) Tehsil
16. In a farm situation for getting firsthand knowledge we use:
    a) Village fairs b) **Farm and home visit**
    c) Telephone d) Questionnaire
17. The **'takes to acceptance'** ratio is the highest in:
    a) **Farm and home visit** b) Method demonstration
    c) Result demonstration d) Campaign
18. Media forum is a methods of:
    a) Individual contact b) **Group contact**
    c) Mass contact d) All of the above
19. In which methods the farmer meets the extension worker at his work place:
    a) Office visit b) Extension campaign
    c) Result demonstration d) **Farmer's call**

20. For reducing tension which extension method that act as a safety valve:

a) Group discussion b) Method demonstration

c) Result demonstration d) Campaign

21. The principle of Method demonstration is:

a) Learning is believing b) Learning by teaching

c) Seeing and doing **d) Learning by doing**

22. Among them which indicates the beliefs:

a) Past knowledge c) Tested empirical Knowledge

c) **Neither true nor false** d) Practical knowledge

23. In a social system conflict in norms occurs when:

a) The members of a group clash with exact other for the exact interpretation of the norm

b) Norms happen to be different for different people

c) There is sub struggle in the society

d) Individuals of different groups tend to confirm to different norms.

24. Cooperative society is basic institution for:

a) Political growth of the rural

b) Social growth of the rural

c) Cultural growth of the rural

d) Socio economic growth of the rural community

25. Culture is a:

a) Limitation on freedom

b) Imitating the individuals in the group ways

c) Whims of the conservative in the group

d) None

26. In India family planning was started in the year:

a) 1952 c) 1954

c) 1956 d) 1958

27. Family is an example of:

a) Primary group c) Secondary group

c) Economic group d) None

28. In a filiocentric family, important role is played by:"

a) Mother c) Father

b) Children d) all

29. The innovation decision process also called as:
    a) **KPDIC**
    b) AIDTCA
    c) AIDCAS
    d) AIETA
30. The stage of innovation decision making at which reinforcement occurs:
    a) Persuasion stage
    b) Knowledge stage
    c) **Confirmation stage**
    d) Decision stage
31. The decision to adopt or reject made after the prior innovation decision is called:
    a) Optional decision
    b) **Contingent decision**
    c) Collective decision
    d) authority decision
32. The fastest rate of adoption of innovation occurs in:
    a) Optional innovation decisions
    b) Collective innovation decisions
    c) **Authority innovation decisions**
    d) Contingent innovation decision
33. The different between decision making unit and adoption unit are:
    a) Optional decision
    b) Collective decision
    c) **Authority decision**
    d) Mental decision
34. An innovation is changed or modified by a user in the process of its adoption is known as:
    a) **Re-invention**
    b) Re-engineering
    c) Re-modeling
    d) Re-making
35. Stage at which an individual forms a favorable or unfavorable attitude towards an innovation:
    a) Confirmation stage
    b) Decision stage
    c) Knowledge stage
    d) **Persuasion stage**
36. The activity when an individual's becomes aware of the objects around one self and events known as:
    a) Participation
    b) **Perception**
    c) Perpetuation
    d) Predetermination
37. Which comes under mass media channel:
    a) Radio
    b) Newspaper
    c) Television
    d) **All of the above**

38. Effect of mass communication is to be consumed immediately hence is called:
    a) Rapid
    b) **Transient**
    c) General
    d) None
39. The term Homophily & Heterophily was coined by:
    a) Rogers
    b) Gabriel Tarde
    c) **Lazersfield and Merton**
    d) None of the above
40. The flow of information through a communication channels is controlled by:
    a) **Gate keeping**
    b) Gate locking
    c) Gate security
    d) Gate guard
41. In a social system established behavior patterns for the members known as:
    a) **Norms**
    b) Values
    c) Attitudc
    d) Beliefs
42. Among the members of a social system patternized social relationship called as:
    a) **Social structure**
    b) Social system
    c) Social organization
    d) Social unit
43. The theory of social change was given by:
    a) D Berlo
    b) **K Lewin**
    c) E M Rogers
    d) P Leaganes
44. According to Kurt Lewin, stages in the social change are:
    a) Unfreezing
    b) Moving
    c) Refreezing
    d) **All of the above**
45. In change analysis, microanalysis approach indicates the:
    a) **Change at individual level**
    b) Change at family level
    c) Change at group level
    d) Change at social system level
46. Primary group is characterized by:
    a) Face to face contact
    b) **Personal and emotional relationship**
    c) Interpersonal proximity
    d) Unity in diversity

47. An in-group is essential character is:
    a) **Intimate relationship** b) Jealousy and competition
    c) Sense of attachment d) Element of sympathy
48. With the help of sociogram we know the:
    a) **Opinion diversification** b) Values
    c) Participation d) None of the above
49. Who actually initiates the action in the community is:
    a) Professional leader b) Assured leader
    c) **Operational leader** d) Popular leader
50. An animal husbandry innovator is known as:
    a) Traditional leader b) Opinion leader
    c) **Functional leader** d) Formal leader
51. An individual acts as opinion leader for a variety of topics:
    a) Monomorphic leadership b) Idiomorphic leadership
    c) **Polymorphic leadership** d) None of the above
52. The decision to make full use of an innovation as best course of action:
    a) **Adoption** b) Diffusion
    c) Innovation d) Technology
53. Adoption process is a:
    a) **Mental process** b) Social process
    c) Psychological process d) Physical process
54. Who say that the stages of adoption are dynamic and not static:
    a) Ryan and Gross b) Wilkening
    c) Johnson d) **Y P Singh**
55. According to Rogers the first stage in the adoption process is:
    a) **Attention** b) Interest
    c) Desire d) Satisfaction
56. Individual decides to continue the full use of the innovation in which stage:
    a) Evaluation b) **Adoption**
    c) Awareness d) Trial
57. At which stage of adoption group discussion is most effective:
    a) Awareness b) **Evaluation**
    c) Testing d) Persuasion

58. Motivating people to change by unfreezing existing behaviour is occurs at:

    a) **Desire stage** b) Interest stage
    c) Conviction stage d) Attention stage

59. The logic of collective action was given by:

    a) R M Rogers b) Gabriel Tarde
    c) Schelling D d) **Mancur Olson**

60. At which stage of adoption poster is most effective:

    a) **Awareness** b) Evaluation
    c) Testing d) Persuasion

61. Example of Symbolic adoption is:

    a) **Covert behaviour** b) Overt behaviour
    c) Affection d) Perception

62. In a social system, time of adoption of an innovation by all members is:

    a) Always same b) **Always vary**
    c) May vary d) All of the above

63. Adopters categories are classified on the basis of:

    a) Mean and variance
    b) **Mean and standard deviation**
    c) Mean and median
    d) Median and standard deviation

64. Among them who considered as deviant by other members of the society:

    a) **Innovators** b) Early majority
    c) Early adopters d) Laggards

65. In which adopter category an opinion leader is belong:

    a) Innovators b) Early majority
    c) **Early adopters** d) Late majority

66. Who adopt an innovation before every average member of the society adopt it:

    a) Innovators b) **Early majority**
    c) Early adopters d) Late adopters

67. 'Legitimization' is done by:
    a) Innovators
    b) **Early Adopters**
    c) Early Majority
    d) Late Majority
68. The percentage of innovators in a social system is:
    a) **2.5 Percent**
    b) 7.5 Percent
    c) 16.5 Percent
    d) 33 Percent
69. An innovator having character of:
    a) **Venturesome**
    b) Skeptical
    c) Traditional
    d) Respectfulness
70. An innovation previously adopt but after some time person decide to reject is known as:
    a) **Active rejection**
    b) Passive rejection
    c) Pro-active rejection
    d) None of the above
71. Discontinuance of an innovation occurs only after:
    a) **Fully adopted**
    b) Partially adopted
    c) Not at all adopted
    d) None of the above
72. For adoption of a better idea, decision to reject an existing idea is known as:
    a) **Disentrancement discontinuance**
    b) Replacement discontinuance
    c) Betterment discontinuance
    d) Distinguishable discontinuance
73. The main economic problem of India are:
    a) Stagnation
    b) Unemployment
    c) Poverty
    d) **All**
74. The term 'Innovation decision process' was given by:
    a) Wilkening
    b) **Rogers**
    c) Johnson and Haver
    d) Ryan and Gross

7

Programmes for Rural Development (Theory)

Rural Development: The economic betterment of people as well as greater social transformation. The efforts for rural development comprises the direct and indirect involvement of people at the grass root level for development of agriculture/animal husbandry, industry, communication, education, health and allied sectors.

Democratic Decentralization: To ensure people's participation in rural development programmes, democratic decentralization has been the core principle of the government since 1957. In the process, people's participation in all developmental activities has been the unique phenomena to find out the solutions of their problems.

Panchayat Raj: On the principle of democratic decentralization, Panchayat Raj institutions were launched in 1957 as per the recommendations of Balwant Rai Mehta Committee. Under this, the government has derived its authority from the people, and distribute to some extent to the people for decision and action at the local level called Panchayat Raj.

Institute Village Linkage Programme (IVLP): IVLP is a programme introduced by ICAR to assess and refine the technologies released by ICAR institutes and State Agricultural Universities (SAUs) at the farmers' field by the scientists together with the practicing farmers.

The 4-H motto is "To make the best better", while its **slogan** is "Learn by doing" sometimes written as "Learn to do by doing".

Five Year Plans

Five-Year Plans (FYPs) are centralized and integrated national economic programs. Joseph Stalin implemented the first FYP in the Soviet Union in the late 1920s. India launched its First FYP in 1951, immediately after independence.

First Plan (1951-1956)

The first five-year plan was launched in 1951 which mainly focused in development of the agricultural sector. The First Five-Year Plan was based on the Harrod–Domar model. The target growth rate was 2.1% annual gross domestic product (GDP) growth; the achieved growth rate was 3.6%.

Second Plan (1956-1961)

The second plan, particularly in the development of the public sector. The plan followed the Mahalanobis model.

Third Plan (1961–1966)

The third five-year plan stressed agriculture and improvement in the production of wheat, but the brief Sino-Indian War of 1962 exposed weaknesses in the economy and shifted the focus towards the defence industry and the Indian Army. Due to miserable failure of the Third Plan the government was forced to declare "plan holidays" (from 1966–67, 1967–68, and 1968–69). Three annual plans were drawn during this intervening period. During 1966-67 there was again the problem of drought. Equal priority was given to agriculture, its allied activities, and industrial sector. The main reasons for plan holidays were the war, lack of resources, and increase in inflation.

Fourth Plan (1969–1974)

At this time Indira Gandhi government nationalised 14 major Indian banks and the Green Revolution in India advanced agriculture. The target growth rate was 5.6%, but the actual growth rate was 3.3%.

Fifth Plan (1974–1979)

The fifth five year plan laid stress on employment, poverty alleviation (Garibi Hatao), and justice. The plan also focused on self-reliance in agricultural production and defense.

Rolling Plan (1978–1980)

The Janta party government rejected the fifth five year plan and introduced a new sixth five year plan (1978-1983). This plan was again rejected by the Indian National Congress government in 1980 and a new sixth plan was made. The earlier one was subsequently referred to as a rolling plan.

Sixth Plan (1980–1985)

The sixth five year plan marked the beginning of economic liberalization. Price controls were eliminated and ration shops were closed. This led to an increase in food prices and an increase in the cost of living. Family planning was also expanded in order to prevent overpopulation.

Seventh Plan (1985–1990)

The plan laid stress on improving the productivity level of industries by upgrading of technology. The main objectives of the seventh five year plan were to establish growth in areas of increasing economic productivity, production of food grains, and generating employment.

Annual Plans (1990-1992)

The eighth plan could not take off in 1990 due to the fast changing political situation at the centre and the years 1990-91 and 1991-92 were treated as annual Plans. .

Eighth Plan (1992–1997)

Modernization of industries was a major highlight of the eighth plan. The major objectives included, controlling population growth, poverty reduction, employment generation, strengthening the infrastructure, institutional building, tourism management, human resource development, involvement of Panchayati raj, Nagar Palikas, NGOs, etc.

Ninth Plan (1997-2002)

The ninth five year plan tried primarily to use the latent and unexplored economic potential of the country to promote economic and social growth. It offered strong support to the social spheres of the country in an effort to achieve the complete elimination of poverty. Special Action Plans (SAPs) were evolved during the ninth five year plan to fulfill targets within the stipulated time with adequate resources. The SAPs covered the areas of social infrastructure, agriculture, information technology and water policy.

Tenth Plan (2002–2007)

The main objectives of the tenth five-year plan were:

- Attain 8% GDP growth per year.
- Reduction of poverty rate by 5% by 2007.
- Providing gainful and high-quality employment at least to the addition to the labour force.
- Reduction in gender gaps in literacy and wage rates by at least 50% by 2007.
- 20-point program was introduced.

Eleventh Plan (2007-2012)

- Empowerment through education and skill development.
- Reduction of gender inequality.
- Environmental sustainability.
- To increase the growth rate in agriculture, industry and services to 4%,10% and 9% respectively.

Twelfth Plan (2012–2017)

The twelfth five year plan of the Government of India has decided for the growth rate at 8.2% but the National Development Council (NDC) on 27 Dec 2012 approved 8% growth rate for 12th five-year plan.

Early Extension Efforts in India

1880	Famine Commission
1920	Gurgaon Experiment (Mr. F.L.Brayne)
1921	Marthandam Attempt (Dr. Spencer Hatch)
1921	Shriniketan Attempt (Shri Rabindra Nath Tagore)
1923	Sevagram Attempt (Shri Mahatma Gandhi)
1928	Royal Commission
1934	Rural Reconstruction Institute (Sir Daniel Hamilton)
1945	India Village Service (Mr. A.T. Mosher & Mr. B.W. Gupta)
1947	Firka development scheme
1947	Grow More Food Campaign
1948	Nilokheri Project or Muzdoor Manzil (Dr. S.K.Dey)
1948	Etawah Pilot Project (Mr. Albert Mayer)
1949	Fiscal Commission by GOI
1951	Bhoodan Movement (Acharya Vinoba Bhave)

Programmes Related to Animal Husbandry Development

1919-20	First census of cattle
1928	Premium Bull Scheme
1946	AMUL (First farmers dairy cooperative)
1964-65	Intensive Cattle Development Programme
1965	National Dairy Development Board
1970 1979-85 1985-96	Operation Flood-I Operation Flood-II Operation Flood-III
1976-77	Special Livestock Breeding Programme
2005-06 to 2007-08	Livestock Insurance Scheme
2000	National project for Cattle and Buffalo Breeding
2005-06	Fodder and Feed Development scheme
2006	Intensive Dairy Development Programme
2006	Intensive Poultry Development Project
2010	Pig Development scheme
2011-12	National mission for protein supplement

National *Extension* Systems

COMMUNITY DEVELOPMENT PROGRAMME			
1952	CDP	Community Development Programme	Overall development of rural areas and people's participation
1953	NES	National Extension Services	Work as an agency to intensify the impact of extension service on farmers
1954	CDB	Community Development Block	Administering rural development programme in rural area
1957	Panchayati Raj	Democratic Decentralization	Local self Government
Development with Social Justice Programme			
1970-71	SFDA	Small Farmers' Development Agency	Technical & financial assistance to small farmers
1970-71	MFAL	Marginal Farmers' and Agricultural Laborers Programme	Technical & financial assistance to marginal farmers
1973	CSRE	Crash Scheme for Rural Employment	For rural employment
1970-71	DPAP	Drought Prone Areas Programme	Protection from drought by achieving environment balance and by developing ground water
1972-73	PPTD	Pilot Project For Tribal Development	Development of six selected tribal areas of AP, MP, Bihar and Orissa
1974	T&V	Training and Visit Programme	World bank assisted programme for TOT
1977	NIRD	National Institution of Rural Development	Training, investigation and advisory for rural development
1977-78	-	Antyodaya Yojna	Scheme of Rajasthan, providing economic assistance to poorest families
1978-79	IRDP	Integrated Rural Development Programme	Assist the BPL families and enable them to earn above poverty line
1979	TRYSEM	Training of Rural Youth for Self Employment	Educational and vocational training
1980	NREP	National Rural Employment Programme	Generation of additional employment for unemployed and under employed person

1982	DWCRA	Development of Women & Children in Rural Areas	Sustainable opportunities of self employment to the women belonging to the rural families who are living below the poverty line
1986	CAPART	Council of Advancement of People's Action & Rural Technology	Assistance to rural people
1983	NAEP	National Agricultural Extension Project	To bridge the gap between well developed research system and extension system
1986	TMO	Technology Mission on Oilseeds	Increase the production & reduce import and achieve self sufficiency in edible oil
1989	JRY	Jawahar Rozgar Yojana	Employment to rural unemployed
1999	-	Annapurna Yojna	10 kgs food grains to elderly people
1993	EAS	Employment Assurance Scheme	100 days assured employment during lean agriculture season
1994	SFAC	Small Farmers Agri-Business Consortium	To build up organization and institute for benefits of farmers from scheme related to agri- business and marketing
1980	DRDA	District Rural Development Agency	Anti poverty programme, financial assistance to rural people by district level authority
1999	SGSY	Swarnjayanti Gram Swarozgar Yojana	Self employment in rural areas
2000	PMGY	Pradhan Mantri Gramodaya Yojna	Basic needs of rural people
2000	AAY	Antyodaya Anna Yojna	To provide food security to poor
2000	PMGSY	Pradhan Mantri Gram Sadak Yojna	Connect all villages with nearest pukka road
2001	SGRY	Sampoorna Grameen Rozgar Yojna	Employment and food security to rural people
2005	NRHM	National Rural Health Mission on remotest areas of the country.	Accessible, affordable, accountable, quality health services to the poorest of the poor

Technological Development Programme			
1960	IADP	Intensive Agricultural District Programme	To provide loan for seeds and fertilizers to farmers
1964	IAAP	Intensive Agricultural Area Programme	To develop special harvest in agriculture are a)
1964-65	ICDP	Intensive Cattle Development Project	Cover up 1 lakhs indigenous cows for cross breeding
1966	HYVP	High Yielding Variety Programme	To increase the productivity of food grains by adopting latest varieties of crops.
1980	IRDP	Integrated Rural Development Programme	overall development of rural poor
1998-99	-	Kisan Credit Card scheme	To provide timely and adequate credit support to the farmers
2004	KCC	Kisan Call Center	KCC agents responding farmer's queries in 23 local dialects. Toll free no.1551, from 6 A.M. to 10 P.M.
1952	KVS	Key Village Scheme	A key village block consisted of one artificial insemination center and four key village units having 500 cows/buffalo
2007-08	RKVY	Rashtriya Krishi Vikas Yojana	Provide incentives to states to achieve 4 % growth in agri and allied sector
INFRASTRUCTURE DEVELOPMENT PROGRAMME			
1999	NATP	National Agricultural Technology Project	With the help of research-extension–farmers linkage provide location specific, demand driven TOT to the farmers
2004	PURA	Providing Urban Amenities in Rural Areas	To achieve balanced socio-economic development in rural areas
2006	NAIP	National Agricultural Innovation Project	To undertake basic and strategic research
2006	NREGA	National Rural Employment Guarantee Act	Provides 100 days of wage employment in a year to every rural household
2005-06	NHM	National Horticulture Mission	To provide holistic growth of horticulture sector

2007	NFSM	National Food Security Mission	National Food Security Mission will have 3 components (i)Rice (ii) Wheat &(iii) Pulses. Restoring soil Fertility and productivity at the individual farm level
2005-06	MoA	Livestock Insurance Scheme (India)	Insurance to cattle and attaining qualitative improvement in livestock and their products.
TRANSFER OF TECHNOLOGY PROJECT OF ICAR			
1964	AICPND	All India Coordinated Project on National Demonstration	Improvement of agricultural crops, but later extended to all aspects of crop and animal husbandry
1974-75	ORP	Operational Research Project	To test the performance of research result on farmers field condition
1974	KVK	Krishi Vigyan Kendra	To provide strong training support to farmers
1979	LLP	Lab to Land Programme	To enhance the socioeconomic condition of farmers by TOT developed by agricultural and veterinary university

Multiple Choice Questions

1. On the basis of an average daily intake, poverty lines in rural area is:
   a) 2100 calories b) 2300 calories
   c) **2400 calories** d) 3400 calories
2. The 'Garibhi Hatao' slogan was given by:
   a) Jawaharlal Nehru b) Mahatma Gandhi
   c) Rabindra Nath Tagore d) **Indira Ghandi**
3. According to 2011 census literacy rate of India is:
   a) 75.85 b) 54.16
   c) 62.67 d) **74.04**
4. Rural development at grass root level occurs by:
   a) **Panchayat** b) Block
   c) District d) State
5. In which year 'DRDA' was established:
   a) 1919 b) **1980**
   c) 1967 d) 1947
6. In which year 'NABARD' was established:
   a) 1935 b) 1959
   c) 1976 d) **1982**
7. The National Commission of Women was set up in the year:
   a) 31st January, 1996 b) **31st January 1992**
   c) 31st January 1995 d) 31st January 1993
8. The ishikawa technique is greatly used to identify:
   a) Input utilization b) **Cause effect relationship**
   c) Policy lacunas d) None
9. In which year 'Ministry of Panchayat Raj' was established:
   a) 27th May 2004 b) **15th August 2002**
   c) 2nd October 2001 d) 15th August 2006
10. Among them those published the 'Journal of Rural Development':
    a) MANAGE b) NAARM
    c) **NIRD** d) NCAP

11. Full form of CAPART is:
    a) Council for Action and People Advancement in Rural Technology
    b) **Council for Advancement of People Action and Rural Technology**
    c) Commission for Action and People Advancement in Rural Technology
    d) Committee for Advancement of People Action and Rural Technology
12. In which year 'Khadi and Village industries Commission' was established:
    a) **1957** b) 1959
    c) 1976 d) 1982
13. The rural development journal is a?
    a) **Kurukshethra** b) Yojanan
    c) Interaction d) Communicator
14. International women's day is:
    a) **8th March** b) 16th October
    c) 1st November d) 22nd December
15. In which year right to information act was passed:
    a) **2005** b) 2006
    c) 2007 d) 2008
16. Among the following which shows the 'Mission 2007':
    a) Medicinal Plants b) Micro irrigation
    c) Plantation sector d) **Knowledge centers**
17. The millennium development goal of reducing world poverty by half is aimed by:
    a) 2030 b) **2015**
    c) 2040 d) 2020
18. Who introduced the zamindari system?
    a) **Mughals** b) British
    c) Marathas d) All of the above
19. In India, "Zamindari" system was abolished in the year:
    a) **1951** b) 1954
    c) 1968 d) 1960
20. 'One Village One Society' principle was given by:
    a) Mehta committee b) **Mec leagan**
    c) Wanchoo d) None

21. The percentage of women covered under JRY:
    a) **30** b) 40
    c) 50 d) 20
22. In which year sriniketan institute was established:
    a) 1915 b) **1914**
    c) 1913 d) 1903
23. At sriniketan rural development was initiated by:
    a) **R N Tagore** b) F L Brayne
    c) S K Dey d) M K Gandhi
24. Sewagram project started by:
    a) R N Tagore b) B P Pant
    c) **M K Gandhi** d) F L Brayne
25. In which year 'Sewagram project' was started:
    a) **1920** b) 1931
    c) 1935 d) 1937
26. In which state 'Gurgaon attempt' was initiate:
    a) Madhya Pradesh b) Uttar Pradesh
    c) **Hariyana** d) Punjab
27. Gurgaon attempt was introduced by:
    a) R N Tagore b) B P Pant
    c) M K Gandhi d) **F L Brayne**
28. In which year 'Gurgaon attempt' was started:
    a) 1918 b) **1920**
    c) 1921 d) 1922
29. "Village Guide" concept was introduced by:
    a) R N Tagore b) B P Pant
    c) M K Gandhi d) **F L Brayne**
30. In which year 'Marthandam Project' was started:
    a) 1918 b) 1944
    c) **1921** d) 1928
31. Person associated with YMCA:
    a) **Spencer Hatch** b) F L Brayne
    c) S K Dey d) Albert Mayor

32. The year in which royal commission on agriculture was came:
    a) 1871 b) 1905
    c) **1928** d) 1929
33. 'Indian Village Service' was started at:
    a) Wardha b) Pohri
    c) Gwalior d) **Lucknow**
34. In which year 'Etawah pilot project' was started:
    a) 1947 b) 1945
    c) 1946 d) **1948**
35. The initiator of Etawah pilot project was:
    a) **Albert Mayor** b) F L Brayne
    c) S K Dey d) M K Gandhi
36. The state of Etawah district is:
    a) Andhra Pradesh b) Bihar
    c) Madhya Pradesh d) **Uttar Pradesh**
37. Nilokheri project was also known as:
    a) **Mazdoor Manzil** b) Kishan Manzil
    c) Mazdoor Madad d) None of these
38. In which year 'Nilokheri project' was started:
    a) **1948** b) 1947
    c) 1950 d) 1951
39. To rehabilitate the refuges from Pakistan during partition which project was started:
    a) Gurgaon project b) **Nilokheri project**
    c) Srinikethan project d) Sewagram project
40. In which year firka development scheme was started:
    a) 1947 b) 1945
    c) **1946** d) 1944
41. Who started the 'Firka development scheme':
    a) R N Tagore b) **T Prakasam**
    c) M K Gandhi d) F L Brayne
42. In which year 'Grow more food' campaign was started:
    a) **1947** b) 1945
    c) 1946 d) 1944

43. GMF enquiry committee was formed in the year:
    a) **1952** b) 1947
    c) 1955 d) 1949
44. Among them who headed the GMF enquiry committee:
    a) **V T Krishnamachari** b) M S Swaminathan
    c) M K Gandhi d) F L Brayne
45. Inventor of sarvodaya scheme was:
    a) R N Tagore b) **Vinoba Bhave**
    c) M K Gandhi d) F L Brayne
46. In which year 'Key village scheme' was started:
    a) 1947 b) 1945
    c) 1946 d) **1952**
47. At sundarban, Rural Reconstruction Institute was started by:
    a) R N Tagore b) **Daniel Hamilton**
    c) M K Gandhi d) F L Brayne
48. In which year 'Bhoodhan' movement was initiated:
    a) 1947 b) 1948
    c) **1951** d) 1963
49. Who was started the bhoodan movement:
    a) **Acharya Vioba Bhave** b) Mahatma Gandhi
    c) Jaya Prakash Narayan d) Jawaharlal Nehru
50. Bhoodan movement was started at:
    a) Warda b) Lucknow
    c) **Pohampallli** d) None
51. Inauguration of CD projects in India was on:
    a) 26th January 1952 b) **2nd October 1952**
    c) 8th August 1954 d) 8th September 1953
52. From which structure of CD blocks were taken:
    a) **Ettawa project** b) Firka project
    c) GMF campaign d) Nilokari experiment
53. Sponsoring agency for CD is:
    a) Government agencies b) Private agencies
    c) Voluntary agencies d) **Both a and b**

54. In India, community development programme was launched as:
    a) Developmental organization   b) Administrative setup
    c) **As means to an end**   d) Local self government
55. In which year 'National extension service' was started:
    a) 1952   b) **1953**
    c) 1954   d) 1955
56. NES was considered as a:
    a) Process   b) **Agency**
    c) Project   d) Plan
57. In which year 'Community Development Blocks' was started:
    a) 1952   b) 1953
    c) **1954**   d) 1955
58. Under CD programme, entire nation was covered by the year:
    a) 1955   b) **1958**
    c) 1962   d) 1979
59. In India, panchayat raj was introduced in the year:
    a) 1950   b) 1952
    c) **1958**   d) 1967
60. The system under panchayat raj is:
    a) Local government system   b) Local administration system
    c) **Local self-government system**   d) None of the above
61. The term Panchayat Raj was coined by:
    a) M K Gandhi   b) **J L Nehru**
    c) B R Mehta   d) M S Mehta
62. The word "Democracy" not indicate except:
    a) Peoples participation   b) **Peoples rule**
    c) Peoples government   d) Peoples judgment
63. In which year constitutional 73rd amendment act was passed:
    a) 1996   b) 2001
    c) **1992**   d) 1997
64. The election to Panchayat is held after a period of:
    a) 3 years   b) 4 years
    c) **5 years**   d) 6 years

65. The Panchayats raj is basically a:
    a) Two tier system
    b) **Three tier system**
    c) Four tier system
    d) One tier system
66. In which state have no Panchayati Raj:
    a) Karnataka
    b) Bihar
    c) Orissa
    d) **Megalaya**
67. States in which 73rd amendment act does not apply:
    a) Nagaland
    b) Mizoram
    c) Megalaya
    d) **All of the above**
68. In panchayat raj institutions, intermediate level is:
    a) Zilla Parishad
    b) Grama Sabha
    c) **Panchayat Samithi**
    d) Block Panchayat
69. In India, invitation to 'Ford foundation' was given on:
    a) 1958
    b) **1959**
    c) 1962
    d) 1979
70. In which year 'IADP' was launched:
    a) 1964 Rabi
    b) 1960 Rabi
    c) 1964 Kharif
    d) **1960 Kharif**
71. The IADP was also known as:
    a) **Package programme**
    b) Training programme
    c) Community programme
    d) Agricultural programme
72. Among them which extension approach used in IADP:
    a) Proscriptive
    b) **Prescriptive**
    c) Participatory
    d) Commercial
73. In which year IAAP was launched:
    a) 1960
    b) 1960
    c) **1964**
    d) 1964
74. In which year 'High Yielding Variety Programme' was started:
    a) 1960
    b) 1960
    c) **1964**
    d) 1976
75. In which year radio rural forum was started:
    a) 1936
    b) 1956
    c) 1958
    d) **1966**

76. In which year broadcasting of rural programme on television was started:
    a) 1964 b) 1965
    c) **1967** d) 1968
77. The Green revolution term was given by:
    a) **William Gadd** b) M S Swaminathan
    c) Norman E Borlaoug d) C Subramaniam
78. In which year 'T & V system' was introduced:
    a) **1974** b) 1978
    c) 1964 d) 1967
79. Sponsoring agency of T & V system was:
    a) WHO b) USAID
    c) **World Bank** d) IMF
80. The T & V system firstly originated in:
    a) India b) Bangladesh
    c) **Turkey** d) Israel
81. Who introduced the T & V system of extension in India:
    a) **Danial Benor** b) Van Den Ban
    c) M S Swaminathan d) C Subramaniam
82. The VLW is also called as:
    a) **VDO** b) VEO
    c) CDO d) VDW
83. The state in which Training and Visit system was introduce:
    a) Punjab **b) Rajasthan**
    c) Tamil Nadu d) Bihar
84. In which year 'Small farmer's development agency' was set up:
    a) 1969-70 b) **1970-71**
    c) 1975-76 d) 1980-81
85. Under SFDA, farmers having Land holding of:
    a) **1-2 hector** b) 0-1 hector
    c) 1-5 hector d) 2-5 hector
86. In which year 'Intensive Cattle Development project' was launched:
    a) 1969-70 b) **1964-65**
    c) 1975-76 d) 1980-81

87. In which year 'Operational flood' was launched:

a) 1986 b) 1981

c) 1976 d) **1971**

88. In which year 'Drought Prone Area Programme' was launched:

a) **1973-74** b) 1968-69

c) 1971-72 d) 1956-57

89. In which year 'Small Farmers Agri-business Consortium' was set up:

a) 1992 b) **1994**

c) 1971 d) 1986

90. In which year 'National Food Security Mission' was launched:

a) 2000 b) **2007**

c) 2008 d) 1986

91. In which year 'Rashtriya Krishi Vikas Yojana (RKVY)' was launched:

a) 2000 b) **2007**

c) 2008 d) 1986

92. In which year, planning commission was setup:

a) 1949 b) **1950**

c) 1951 d) 1952

93. The chairman of planning commission was:

a) Presidentes b) Speaker

c) **Prime Ministre** d) None

94. The Planning commission is an:

a) Statutory body b) **Advisory body**

c) Autonomous body d) Elected body

95. The planning commission of India is a:

a) **Constitutional body** b) Statutory body

c) Non-statutory body d) Autonomous body

96. Among them all are member of National Development Council except:

a) Primer Minister b) **President**

c) Chief Ministers d) Cabinet Ministers

97. The national development council was set up in the year:

a) 1950 b) 1951

c) **1952** d) 1953

98. The programme evaluation organization (PEO) in integral part of the:
    a) National Development Council
    b) Central Secretariat
    c) Department of Public Enterprises
    d) **Planning Commission**
99. The plan implementation and its impact evaluation is done by:
    a) NDC
    b) Union Cabinet
    c) **PEO**
    d) Planning Commission
100. The first five year plan was submitted for approval in the year:
    a) 1950
    b) **1951**
    c) 1952
    d) 1953
101. In which year 'IRDP' was introduced:
    a) 1975-76
    b) **1978-79**
    c) 1980-81
    d) 1981-82
102. The approach followed in IRDP is:
    a) Individual based
    b) Group based
    c) Community based
    d) **Family based**
103. IRDP is an:
    a) Adult education programme
    b) **Anti-poverty programme**
    c) Anti-boarding programme
    d) Fisheries programme
104. Under IRDP, basic unit of the development is:
    a) Farmers
    b) Village
    c) **CD block**
    d) Family
105. DRDA is established at:
    a) State level
    b) **District level**
    c) Tehsil level
    d) Village level
106. In which year 'TRYSEM' was introduced:
    a) 1989
    b) **1979**
    c) 1969
    d) 1978
107. DWCRA was implemented on:
    a) **1982**
    b) 1979
    c) 1974
    d) 1978

108. The skill requirement of the DWCRA members was achieved by:

a) SFDA | b) IRDP
c) NREP | d) **TRYSEM**

109. In which year 'Swarnajayanti Gram Swarozgar Yojana (SGSY)' was launched:

a) **April 1999** | b) October 1993
c) November 2001 | d) March 1987

110. SGSY was formed by integrating the programmes:

a) SITRA | b) GKY
c) MWS | d) **All**

111. The share of state and center in the funding of SGSY is:

a) **25:75** | b) 60:40
c) 50:50 | d) 25:75

112. For beneficiary of SGSY, minimum net monthly earning is:

a) Rs. 1000 | **b) Rs. 2000**
c) Rs. 2500 | d) Rs. 3500

113. The SHG formed under SGSY should have BPL families:

a) **80 percent** | b) 63.33 percent
c) 50 percent | d) 40 percent

114. Under SGSY, maximum subsidy provided for individuals:

a) 5000 | b) **7500**
c) 2000 | d) 9000

115. Bank credit under SGSY is available on group up to:

a) 3 times | b) **4 times**
c) 5 times | d) 6 times

116. In which year 'Swarna Jayanti Shahari Rozgar Yojana (SGSRY)' was launched:

a) **1st December 1997** | b) 2nd October 1995
c) 14th November 2001 | d) 2nd October 2001

117. Antyodaya Anna Yojana was launched in the year:

a) 1999 | b) **2000**
c) 2001 | d) 2002

118. Up to which level mid day meal scheme covers the children:
    a) **Primary level**
    b) Secondary level
    c) High school level
    d) College level
119. The Mid Day Meal Scheme was started in the year:
    a) **1995**
    b) 2000
    c) 2001
    d) 2005
120. In which year 'Annapurna Scheme' was started:
    a) 1999
    b) 2004
    c) **2000**
    d) 2005
121. In which year 'Food for work programme' was started:
    a) **1977**
    b) 1980
    c) 1981
    d) 1982
122. NREP was started in the year:
    a) 1977
    b) **1980**
    c) 1981
    d) 1982
123. NREP was check:
    a) **Unemployment**
    b) Population
    c) Brain Drain
    d) Drought
124. JRY was started in the year:
    a) 1977
    b) **1989**
    c) 1981
    d) 1982
125. In which year 'Jawahar Gram Samridhi Yojana (JGSY)' was launched:
    a) **April, 1999**
    b) October, 1993
    c) November, 2001
    d) March, 1987
126. At which level 'Jawahar Gram Samridhi Yojana' is implemented:
    a) District
    b) Intermediate
    c) **Village**
    d) Nagar
127. Sampoorna Grameen Rozghar Yojana (SGRY) was started in the year:
    a) 1999
    b) 1993
    c) **2001**
    d) 1987
128. The national rural employment guarantee act was passed in the year:
    a) 1992
    b) 1996
    c) 1997
    d) **2005**

129. In which year 'Prime Minister's Rozgar Yojana (PMRY)' was started:

a) **1993** b) 1995

c) 1997 d) 2001

130. Under PMRY, which age group get subsidy:

a) **18-35** b) 20-40

c) 15-35 d) 25-50

131. PMRY beneficiary must have minimum level of education for eligibility is:

a) IV pass b) **VIII pass**

c) X pass d) XII pass

132. Pradhan Manthri Gram Sadak Yojanay was launched during:

a) **December 2000** b) August 2002

c) October 2001 d) August 2006

133. The Pradhan Mantri Gramodya Yojana is related to:

a) Rural sanitation b) Rural water supply

c) **Rural housing** d) Primary education

134. IVLP expand as:

a) International village link programme

b) Institute village link programme

c) **Institutue village linkage programme**

d) Institute village link project

135. Concept of IVLP is:

a) Agro-system analysis b) **Agro-system analysis**

c) Eco-system analysis d) Socio-Ecosystem analysis

136. The bio-village concept was given by:

a) **M S Swaminathan** b) G Perumal

c) Curt Lewin d) Anil Gupta

139. ATMA indicates as:

a) Agency for Training and Management in Agriculture

b) **Agricultural Technology and Management Agency**

c) Assessment Transfer and Monitoring Agency

d) Assessment Training and Management Agency

140. ATMA is an important component of:
a) NARS b) **NATP**
c) NAEP d) KVK

141. Level at which ATMA operates::
a) Block level b) **District level**
c) State level d) National level

142. The number of ATMA developed under NATP:
a) 22 b) 24
c) 25 d) **28**

143. The ATMA is a:
a) Statutory body b) **Registered society**
c) Cooperative society d) Krishi Vigyan Kendra

144. Under ATMA, farmers information and advisory centers operates at:
a) District level b) **Block level**
c) Tehsil level d) Village level

145. The state level apex training institute to give training support to ATMA:
a) MANAGE b) EEI
c) CAPART d) **SAMETI**

146. SREP is an component of:
a) NARS b) **ATMA**
c) ATIC d) KVK

147. Full form of ATIC is:
a) Agricultural Technology Initiation Center
b) Agricultural Technology Information Committee
c) **Agricultural Technology Information Center**
d) Agricultural Technology Invitation Center

148. ATIC functions under:
a) State Department of Agriculture b) Ministry of Agriculture, GOI
c) Ministry of RD, GOI d) **ICAR**

149. Full form of AKIS is:
a) Agricultural Knowledge and Information Syndrome
b) Agricultural Knowledge and Information Service
c) Agriculture Knowledge and Information Science
d) **Agriculture Knowledge and Information System**

150. Full form of MANAGE is:
   a) **National Institute of Agricultural Extension Management**
   b) Management Institute for Agricultural Extension
   c) Institute of Agricultural Management
   d) None of the above

151. Full form of 'MANAGE' was established:
   a) **1986** b) 1987
   c) 1988 d) 1989

152. Under ICAR, first transfer of technology (TOT) project was:
   a) LLP b) ORP
   c) **AICPND** d) KVK

153. The operational research project was launched in the year:
   a) **1974** b) 1979
   c) 1964 d) 1969

154. On Golden Jubilee celebration of ICAR's which programme was launched:
   a) **Lab to land** b) KVK
   c) ORP d) None

155. The relatively non-mobile TOT programmes of ICAR is:
   a) LLP b) ORP
   c) AICPND d) **KVK**

156. In which year NARP was implemented:
   a) **1980** b) 1982
   c) 1984 d) 1986

157. Funding agency of NATP is:
   a) ICAR b) FAO
   c) **World bank** d) ICAR

158. Cooperative extension service is a consequence of:
   a) **Smith Liver act** b) Morill act
   c) Hatch act d) Extension act

159. In which year 'Smith liver act' was passed:
   a) 1862 b) 1891
   c) 1908 d) **1914**

160. Under which act 'Land Grant Colleges' come into existence:

a) Smith Liver act
b) **Morill act**
c) Hatch act
d) Extension act

161. In which year 'Morill act' was passed:

a) **1862**
b) 1891
c) 1908
d) 1914

162. Under which act 'Experimental stations' come into existence:

a) Smith liver act
b) Morill act
c) **Hatch act**
d) Extension act

163. In which year 'Hatch act' was passed:

a) 1862
b) **1887**
c) 1908
d) 1914

164. From which country 'Comilla project' was originated:

a) Bhutan
b) Nepal
c) USA
d) **Bangladesh**

165. Who among the following identified three levels of educational objective?

a) Thompson
b) Thorndike
c) Leagan
d) Max Weber

166. Which characteristic is associated with Planning, Programming and Budgetting System (PPBS)?

a) Emphasis on output
b) Emphasis on input
c) Emphasis on resources
d) None of these

167. The activity which is having preceding activity but no succeeding activity is called __________

a) Bumper
b) Dangler
c) Dumper
d) Ocular

168. Correct expansion of PANR is __________

a) Participatory Assessment of Natural Resources
b) Participatory Analysis of Natural Resources
c) Participatory Analysis of National Resources
d) Participatory Analysis of Normal Rainfall

169. Which of the following was not given due importance in the ICAR classification of 20 Agro-Ecological regions and 60 sub regions?

a) Socio-economic endowment

b) Market support

c) Service sector in agricultural development

d) All of these

170. Which of the following denote a strategic planning tool, used for giving Terms of reference for new projects?

a) BCWP **b) BOSCARD**

c) BCWS d) None of these

171. National Institute for Transforming India (NITI) Aayog was formed on ____

a) 1st January 2015 b) 1st January 2014

c) 15th August 2014 d) 2nt October 2014

172. 'Plan-Do-Check-Act cycle' which is a four-stage approach of problem-solving is also known as

a) Deming cycle b) Shewhart cycle

c) Omnibus cycle **d) Both a & b**

173. 'Fishbone diagram' which is also known as cause & effect diagram, has been developed by

a) Alexander Deming **b) Kaoru Ishikawa**

c) Mc Gregor d) Daniel Katz

174. Who has given the theory of 'zero defects'?

a) Mc Callaghan **b) Philip Crosby**

c) Rupert Mc Lloys d) None of these

8

Management in Extension (Theory)

The term 'Management' derived from word 'manage' derived from French word menage means 'housekeeping'.

According to Knootz and Weihrich 1988, 'Management as the process of designing and maintaining an environment in which individuals, working together in groups, efficiently accomplish selected aims'.

According to Louis E.Boone and David L.Kurtz 'management as the use of people and other resources to accomplish objectives'.

POSDCORB is an acronym widely used in the field of Management and Public Administration., it first appeared in a 1937 staff paper by Luther Gulick and Lyndall Urwick. The acronym stands for steps in the administrative process: Planning, Organizing, Staffing, Directing, Coordinating, Reporting and Budgeting.

Planning: Planning can be defined as the process, by which, managers decide the mission and objectives of the firm and take necessary steps to achieve the desired objectives. At the same time, managers need to determine the future trends in business and incorporate change and innovation into the organization from time to time.

Organizing: Organizing is the continuous process of assigning tasks and allocating resources to individuals to enable them to accomplish organization goals.

Staffing: Staffing is better known as "human resource management" and involves manning or filling the various positions in the organizational hierarchy. This function also deals with compensation, performance appraisal, promotion and career planning.

Directing: Directing is the continuous task of making decisions and embodying them in specific and general orders and instructions and serving as the leader of the enterprise.

Co-ordinating: Coordinating is the all-important duty of interrelating the various parts of the work.

Reporting: Reporting is keeping those to whom the executive is responsible informed as to what is going on, which thus includes keeping himself and his subordinates informed through records, research and inspections.

Budgeting: Budgeting, with all that goes with budgeting in the form of fiscal planning, accounting and control.

Controlling: Controlling can be defined as the continuous measurement and analysis of actual operations against the established standards developed during the planning process and corrections of deviations, if any.

Leading: Leading is defined as "the management function of influencing, motivating, and directing people towards the achievement of organizational goals."

Fayol's Fourteen Principles of Management

- Division of Work
- Authority and Responsibility
- Discipline
- Unity of Command
- Unity of Direction
- Subordination of Individual Interest to General Interest
- Remuneration of Personnel
- Centralization
- Scalar Chain (line of authority with peer level communication)
- Order
- Equity
- Stability of Tenure of Personnel
- Initiative

Scientific theory of management: Developed by Frederick Winslow Taylor in 1895.

Key Communicator or opinion leaders are persons in any social system who are sought out for information and advice on general or specific topics. They have been also called as fashion leaders, influence, information leaders, opinion leaders, spark plugs, style-setters, taste makers etc. Three methods are in use for identifying key-communicators i.e.

1. Sociometric method
2. Self-designation technique
3. Information ratings

Unity of command: Also under coordination, as well as organization, Gulick emphasizes the theory of unity of command, that each worker should only have one direct superior so as to avoid confusion and inefficiency.

Span of management: There is a limit to the number of subordinates a manager can effectively supervise. Gulick notes that any manager will have a finite amount of time and energy, and discusses span of control under coordination.

Scalar chain: This refers to the chain of authority that extends from the top to the bottom of an organization. The scalar chain defines the communication path in an organization.

Line and staff function: Line functions are those which have direct responsibility for accomplishing the objectives of the enterprise, and staff refers to those elements of the organization that help the line to work more effectively in accomplishing the primary objectives of the enterprise."

Decentralization & Centralization

According to Louis A. Allen, Centralization is the systematic and consistent reservation of authority at central points within an organization. Decentralization applies to the systematic delegation of authority in an organization-wide context".

According to Robert Kreitner, "Centralization is the relative retention of decision making authority by top management. Decentralization is the granting of decision- making authority by management to lower level employees."

Delegation of Authority: It is the delivery by one individual to another of the right to act, to make decisions, to requisition resources and to perform other tasks in order to fulfill job responsibilities." In other words, individuals need to be given a degree of authority to enable them to carry out their tasks.

Recruiting and Selecting: Recruiting is the process of generating a sufficiently large group of applicants from which to select qualified individuals for available jobs. They are closely related. Their distinction is that recruiting means "searching for and obtaining applicants to hire. Selection, in particular is a complex area, because even the best concepts and techniques will not guarantee that every individual hired will perform well, or that an organization will never be charged with unfair employment practices.

Human Resource Management (HRM): HRM is a management function that helps managers plan, recruit, select, train, develop, remunerate and maintain organization members. HRM is the latest nomenclature used to denote personnel management. HRM has four objectives, viz., (i) societal objectives, (ii) organizational objectives, (iii) functional objectives, (iv) personal objective.

Human Resource Development: Human resource development process is denoted as the means by which personnel are recruited, trained and put into use depending upon their skills, knowledge and potential as per the need of the job and keeping in line with organizational objectives.

Human Resource Planning: To utilize the human resources to its full potential, planning is a must. If planning is effective, meaningful and strategically right, every personnel of the organization would be able to contribute much in their endeavor.

Job Analysis: The procedure for determining the duties and skill requirements of a job and the kind of person who should be hired for it.

Job Evaluation: A systematic comparison done in order to determine the worth of one job relative to another.

Graphic Rating Scale: A scale that lists a number of traits and a range of performance for each. The employee is then rated by identifying the score that best describes his or her level of performance for each trait.

Paired Comparison Method: Ranking employees by making a chart of all possible pairs of the employees for each trait and indicating which the better employee of the pair is.

Performance Analysis: Verifying that there is a performance deficiency and determining whether that deficiency should be corrected through training or through some other means.

Performance Appraisal: Evaluating an employee's current and or past performance relative to his or her performance standards.

Performance Management: Managing all elements of the organizational process that affect how well employees perform, the process followed by the employers to make sure that employees are working towards organizational goals.

Point Method: The job evaluation method in which a number of compensable factors are identified and then the degree to which each of these factors is present on the job is determined.

SWOT Analysis (Strengths, Weaknesses, Opportunities, and Threats): SWOT analysis is carried out in respect of different farming systems as observed in each prevailing agro-ecological situations.

Stages of group formation:

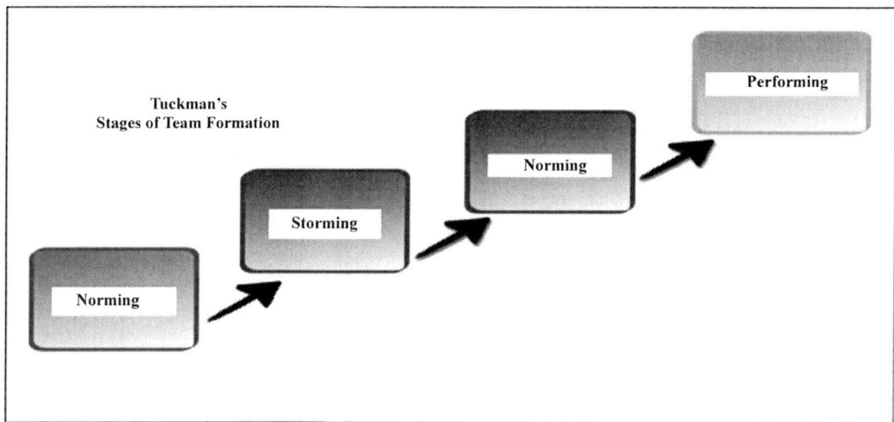

1. **Forming:** In this stage group members enroll themselves and conduct 1-2 initial meetings.
2. **Storming:** In this stage group members start discussing and reacting to the various issues / conflicts
3. **Norming:** The groups start framing norms to run it successful
4. **Performing:** The group starts performing by lending money and collecting it back, involving in developmental programmes, management of inputs and output, marketing, etc.

SQ3R Methods: SQRRR is a reading comprehension method named for its five steps i.e. survey, question, read, recite, and review. The method was introduced by Francis Pleasant Robinson in 1946.

Training Cycle: Training helps in building capacities among the individual so as to improve their performance in their job activities.

Three kinds of training cycle are:

1. Induction training
2. Orientation
3. On the job training

The cycle followed for each kind of training are - planning, implementation and evaluation.

Training Needs Assessment: This is required to find out and identify the needs of individual trainee on which they should build their professional competencies to carry out the assigned job in his organizations. This needs assessment occurs at two levels – groups and individuals.

Curriculum Development: Curriculum Development provides us the framework and foundation for training. The curriculum specifies what will be taught how it will be taught.

Training Needs Analysis: Training Needs Analysis helps to identify the gap which helps to inform the planning of training. It guides the formulation of training objectives and assists in the identification of training contents.

On the Job Training Methods: The most common or formal on the job training programme is training for specific job. There are following methods of training:

i. Experience: This is the oldest method of on-the- job training. But as a sole approach, it is wasteful, time consuming and inefficient.

ii. Coaching-On-the-job: The technique involves direct personnel instruction and guidance, usually, with extensive demonstration and continuous critical evaluation and correction by a superior.

iii. Understudy: The advantage of this method is that training is conducted in a practical and realistic situation.

iv. Position Rotation: The major objective of job rotation training is the broadening of the background of trainee in the organisation.

v. Special Projects: This is a very flexible training device. Such special project assignments grow ordinarily out of an individual analysis of weaknesses.

vi. Selective Reading: Individuals in the organisation can gather and advance their knowledge and background through selective reading of professional journals and books.

vii. Apprenticeship: These arrangements usually provide a mixed programme of classroom and job experience.

viii. Vestibule Schools: A preliminary to actual job experience. Vestibule schools are widely used in training for clerical and office jobs as well as for factory production jobs. Such training is usually shorter and less complex than that adaptable to the apprenticeship system.

Off-the-job Training Methods: In these methods, trainees have to leave their workplace and devote their entire time to the development objective. Following training techniques are used off-the-job:

i. Special course and lectures: Lecturing is the most traditional form of formal training method where the trainer is active and the trainee is passive. The first such programme was the sloan fellowship programme, established in 1931 at Massachusetts institute of technology, U.S.A.

ii. Conferences: In this method, the participants pool their ideas and experience in attempting to arrive at improved methods of dealing with the problems which are common subject of discussion). As the word conference literally means it is a large group of delegates that confer. It is a step ahead of seminar where the number of delegates or participants is comparatively less.

iii. Case studies: This technique developed by the Harvard Business School, U.S.A. is one of the most common form of training. A case is a written account of a trained reporter or analyst seeking to describe an actual situation.

iv. Brainstorming: This is the method of stimulating trainees to creative thinking. This approach developed by Alex Osborn seeks to reduce inhibiting forces by providing for a maximum of group participation and a minimum of criticism. The purpose is to promote group creativity and should be restricted to the twenty people per group. Every brainstorming session can be divided into two, separate stages, namely:

- The green stage: In this stage, all members of a brainstorming session express any ideas that come to their mind in reference to a given task. It is unimportant whether or not these ideas are rational or justified. In the green stage, nobody should evaluate others' ideas in any way.
- The red stage: Participants of a brainstorming session analyze/evaluate all of the previously mentioned ideas and choose those which will

most probably lead to the best possible solutions of a given problem at a given time.

vi. Laboratory Training: Laboratory training adds to conventional training by providing situations in which the trainees themselves experience through their own interaction some of the conditions they are talking about. Laboratory training is more concerned about changing individual behavior and attitude. Laboratory training classified in to two categories i.e.

a) Simulation: In this method, instead of taking participants into the field can be simulated in the training session itself. Simulation is the presentation of real situation of organizations in the training session. There are two common simulation methods of training: role-playing and business game.

i) Role-Playing: Role-Playing is laboratory method which can be used rather easily as a supplement to conventional training methods. Its purpose is to increase the trainee's skill in dealing with other people.

ii) Gaming: Gaming has been devised to simulate the problems of running a company or even a particular department. It has been used at all levels, from the lop executives to the production supervisors. Gaming is a laboratory method in which role-playing exists but its difference is that it focuses attention on administrative problems, while role-playing tends to emphasize mostly feeling and tone between people in interaction.

b) Sensitivity Training: Sensitivity training is a small-group interaction under stress in an unstructured encounter group which requires people to become sensitive to one another's feelings in order to develop reasonable group activity. As a result of criticism and experience, a somewhat revised approach, often described as 'team development' training, was developed. It was first used by National Training Laboratories at Bethel, U.S.A. The training groups themselves called 'T Group' (from ten to twenty members).

Buzz group: Buzz groups are formed by dividing large groups into small discussion groups of 2 to 15 people who meet simultaneously for a specified time to discuss a specific question, problem, or issue. This type of discussion creates independent and cognitive thinking among group members. The groups work simultaneously in the same room (the word buzz comes from the resulting noise). The task is brief and relatively simple.

Variations of the Buzz group:

1. Phillips 66
2. Clark's 22
3. Huddle methods
4. Circular response methods
5. Progressive Buzz session

Panel Discussion: Panel discussion method is one of the training methods when a small group of panelists presents divergent views on some issues worth deliberating. The participants listen and then involve themselves in the combined discussion. Panel is a group of three or more people often skilled in various fields, who conduct before an audience a discussion on a topic in order to stimulate interest and to present different points of views rather than to arrive at a single solution or to establish the superiority of one view.

Colloquium: Modified version of panel discussion in which three or four resource persons discuss a specific topic. In colloquium there is more of audience participation.

Symposium: It is a meeting in which small number of resource persons present paper on a given topic. Each one speaks for a definite time. Symposium is mainly meant for information gathering at the professional level.

Syndicate: It is a study group, has a chairperson and a secretary. A problem is assigned to the syndicate for carefully study and solution.

Workshop: One of the training methods used in management training and technical training is workshop method. It involves practical work done by the participants themselves.

Motivation: Means movement or motion, an inner state that energizes, activates or moves and directs human behavior towards goals. It may generate at two levels i.e. intrinsic motivation (within one's own self) and extrinsic motivation (artificially induced incentives or outside sources).

Theories of Motivation

A) Early theories of motivation

1. **Hierarchy of Needs Theory:** Hierarchy of Needs Theory Abraham Maslow (1943) a humanistic psychologist, propounded a theory of human needs that has widely influenced modern understanding of motivation. He hypothesized that within every human being there exists a hierarchy of five needs.

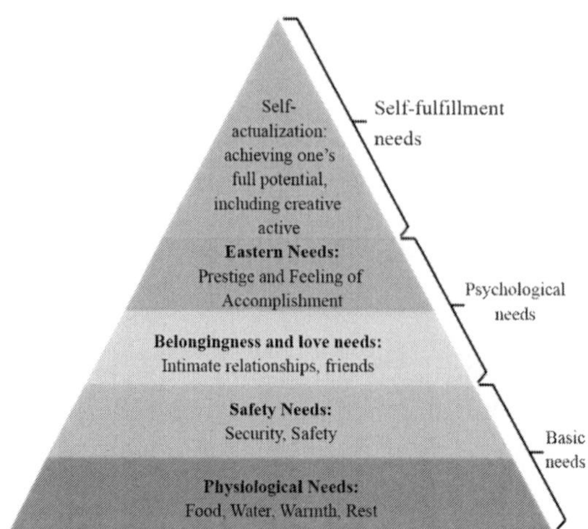

2. **Theory of X & Y:** Douglas Mc Gregor proposed two distinct views of human beings: one basically negative, labeled theory X, and the other basically positive, labeled as Y. After viewing the way managers dealt with employees. Mc Gregor concluded that a manager's view of the nature of (human beings is based on a certain grouping of assumptions, and that he or she tends to hold his or her behaviour towards subordinates according to these assumptions.

Mc Gregor assumptions of theory X

1. Employees inherently dislike work and wherever possible will attempt to avoid it.
2. Since employees dislike work, they must be coerced, controlled or threatened with punishment to achieve desired goals.
3. Employees will shirk responsibilities and seek formal direction whenever possible.
4. Most workers place security above all other factors associated with work, and will display little ambition.

Mc Gregor assumptions of theory Y

1. Employees can view work as being as natural as rest or play.
2. A person will exercise self direction and self control if he is committed to the objectives.
3. The average person can learn to accept, even seek, responsibility.

Mc Gregor, himself, held to the belief that theory Y assumptions were more valid than theory X. Therefore, he proposed ideas like participation in decision making, responsible and challenging jobs, and good group relation as approaches that would maximize and employee's job motivation.

3. **Motivation-Hygiene Theory:** Frederick Herzberg found that people have two different categories of needs in the work situation, and these are essentially independent of each other and affect behaviour in different ways. He found that when people felt dissatisfied with their jobs, they were concerned about the environment in which they were working. On the other hand, when people felt good about their jobs, this had to go with the work itself. Herzberg called the first category of needs hygiene or maintenance factors. Hygiene because they describe people's environment and serve the primary function of preventing job satisfaction, maintenance because they are never completely satisfied. They have to continue to be maintained. He called the second category of needs motivators since they seemed to be effective in motivating people to superior performance.
4. **ERG motivation theory:** Clayton P. Alderfer's (1969) condenses Maslow's five human needs into three categories: **Existence**, **Relatedness** and **Growth** (on the basis of categories' concreteness).

Existence Needs: Include all material and physiological desires (e.g., food, water, air, clothing, safety, physical love and affection). Maslow's first two levels.

Relatedness Needs: Encompass social and external esteem; relationships with significant others like family, friends, co-workers and employers. Maslow's third and fourth levels.

Growth Needs: Internal esteem and self actualization; these impel a person to make creative or productive effects on himself and the environment (e.g., to progress toward one's ideal self). Maslow's fourth and fifth levels.

Existence needs are the most concrete, and easiest to verify. **Relatedness needs** are less concrete than existence needs, which depend on a relationship between two or more people. **Growth needs** are the least concrete in that their specific objectives depend on the uniqueness of each person.

B) Contemporary theories of motivation

1. **Three Needs Theory:** David Mccle II has proposed that there are three major relevant motives or needs in the workplace.
   a) The need for achievement (n Ach) the drive to excel, to achieve in relation to a set of standards, to strive to succeed.
   b) The need for power (n Prw) the need to make others behave in a way that they would not have behaved otherwise.

c) The need for affiliation (n Aff) the desire for friendly and close interpersonal relationships.

2. **Goal Setting Theory:** Goals-can be a major source of work motivation. If people participate in goal setting, they are more likely to accept even a difficult goal than if it is arbitrarily assigned to them by their boss. Thus goal setting theory deals with people in general.
3. **Equity Theory:** The equity theory of motivation formulated by Adams, (1963) focuses on the employee's perception of fair treatment meted out to him/her as an important determinant of his on her level of motivation.
4. **Expectancy Theory:** Victor Verson (1964) formulated a theory which attempts to explain the willingness of an individual to contribute his personal efforts to achieve organizational tasks. The theory is based on the promise that, people will do what they can do when they want to".
5. **Reinforcement Theory:** B.F. Skinner's reinforcement theory considers human behaviour as being largely determined by its consequences. Reinforcements refer to the outcomes of behavior. Outcomes may be desirable to the individual (rewards), or undesirable (punishments).

Leadership: Defined as "the process of influencing the activities of an individual or a group in efforts towards the achievement of goals in a given situation".

Leadership Styles: 1. Autocratic leadership 2. Democratic leadership 3. Laissez-faire leadership

Adam Smith was a Scottish philosopher who coined the term "Laissez-Faire" - Meaning "let do" in French. The laissez-faire leadership style is where all the rights and power to make decisions is fully given to the worker**.** The laissez-faire style is sometimes described as a "hands off" leadership style because the leader delegates the tasks to their followers while providing little or no direction to the followers.

Theories of Leadership

1. **Trait Theory/Great Man Theory:** According to this school of thought, leadership is rooted in biology. Researchers mainly focus on isolating physical and psychological qualities of leaders such as height, strength, body type, enthusiasm, intelligence, self-confidence, decisiveness, bravery and charisma. Some of those who support the trait theory believe that leaders are born, not made, while others take the view that everyone can lead, but he or she will lead differently.

2. **Contingency Theories:** In 1964 Fred Fiedler published the Fiedler Contingency Model of leadership that recognized that the style of leadership that was most effective depended upon the context in which the style was applied. Two major approaches stem from contingency theory i.e. styles and roles.

   The Styles and Roles Approach: This approach analyses the leadership roles and behavioural styles of successful leaders. It acknowledges that group situations are often different, and emphasizes the need for flexibility in styles and roles. A person must play many roles and employ different styles in order to be an effective leader.

3. **Situational Leadership Theory:** This is based on the idea that leadership style should vary with the maturity of the group. It means that task behaviour and relationship behaviour, two critical dimensions of leadership, should change with the level of group development.

4. **Organizational Theory:** According to this theory, leadership is a function of position and role in a hierarchical organization. Some writers make a distinction between managing and leading. To manage means "to bring about, to accomplish, to have charge of or responsibility for, to conduct". Leading is "influencing, guiding in direction, course, action, opinion". The distinction is crucial. Managers are people who do things right, whereas leaders are people who do the right thing. Leadership in this view refers to the head of some group or organization.

5. **Power Theory:** Leadership as power involves the ability to make something happen to satisfy individual, group or task needs. It means being able to make a difference, e.g., getting something done or influencing the direction the group takes. According to one view, this power emanates from positional power (a symbolic position such as President, Director etc.) or operational power (by people who get things done without any formal position).

6. **Vision Theory:** According to this theory, the critical ingredient in leadership is vision. Leaders scan current trends and future threats and opportunities, then engage in strategic redirection of people/groups towards a desired future.

7. **Ethical Assessment Theory**: Leadership, in this view, is inherently ethical. The view is that ethics is at the centre of every human action, and hence, at the centre of leadership, leader-follower relationships and in the leadership's vision of human needs.

8. **Transactional Leadership:** Leaders approach others with a view to exchanging one thing for another, such as jobs for votes, or recognition for services etc.

9. **Transformational leaders:** Leaders motivate their followers, not by exchanging goods, but by sharply arousing or altering the strength of needs that may have lain dormant, e.g., higher order needs for self-actualization.

Likert's management systems: Management styles developed by Rensis Likert in the 1960s. They describe the four systems of management to describe the relationship, involvement, and roles of managers and subordinates in industrial settings. Four management systems are:

1. Exploitative authoritative (System-I): Exploitative authoritative is rooted in classical theory. In this system, managers tend to use threats, fear, and punishment to motivate their workers.
2. Benevolent authoritative (System-II): Under this system motivation is based on the potential for punishment and partially on reward
3. Consultative system (System-III): This theory is very closely related to the human-relations theory. Motivation of workers is gained through rewards, occasional punishments, and very little involvement in making decisions and goals. Communication in this system flows both downward and upward, though upward is more limited.
4. Participative system (System-IV): Likert argued that the participative system was the most effective form of management. This system coincides with human-resources theory. This system promotes genuine participation in making decisions and setting goals through free-flowing horizontal communication and tapping into the creativity and skills of workers.

Organizational conflict: Conflict is generally perceived as dysfunctional.

Styles of managing conflict: There are five basic styles which can be used in managing conflict:

1. Competition
2. Problem-solving
3. Avoidance
4. Accommodation
5. Compromise

Steps in capacity building

1. Enabling
2. Empowering
3. Decision – Making
4. Self-Reliance

Experiencing learning cycles: Kolb (1984): Related with the adult learning process. This suggests that there are four stages in learning which follow from each other:

1. Concrete Experience
2. Reflection
3. Abstract Conceptualisation
4. Active Experimentation

All this may happen in a flash, or over days, weeks or months, depending on the topic, and there may be a "wheels within wheels" process at the same time.

Objectives in a project

SMART objectives need to be set for the project, SMART stands for

Specific: The objective should specify what intends to be achieved.

Measurable: Standards need to be defined so that achievement of objectives can be judged

Achievable: At the field situation, the set objectives should be achievable and realistic

Relevant: Objectives need to be relevant to the needs of local people

Time Bound: Time dimension needs to be defined for achieving the objectives

- **Resources are INPUTS:** Input is good, funds, services, manpower, technology
- **Results are OUTCOMES**: Results are grouped into three broad categories:
    - i) Output: Specific products or services which activity is expected to produce from its inputs in order to achieve the set objectives. (Increased irrigation, fertilizer use, health facility etc.)
    - ii) Effects: Outcomes of the use of Project inputs – (incremental yields, income, etc.)
    - iii) Impact: Outcome of Project Effects (Standard of living and reducing poverty) both at individual and community level.

Project: An activity undertaken with a specific objective involving a specific time frame with a definite beginning and a definite completion.

Project Management: Project management involves decision making for the planning, organizing, coordination, monitoring and control of a number of interrelated time bound activities.

Programme Projection: - 1955 by Extension Committee on organization &

policy (ECOP). According to Lawrence 1962 'It is a long range programme planning by a broadly representative group of local people.

FSR/E: - The Farming system research considers the whole farm as a system with interdependent sub systems such as house hold, crop & animals.

Steps of Farming System Research

1. Diagnostic stage
2. Design stage
3. Testing stage
4. Dissemination stage

Project Life Cycle: The phases of conception, definition, execution and operation, which every project passes through various phases of life cycles.

Economic Rate of Return (ERR): A measure for deciding the economic benefits of the project to the society/economy.

Risk: A situation where the likelihood of a particular outcome can be quantified and risk indicate adversity

Project Formulation: "Project Formulation" is the processes of presenting a project idea in a form in which it can be subjected to comparative appraisals for the purpose of determining in definitive terms the priority that should be attached to a project under severe resource constraints.

Pay-back Period: This is the simplest of all methods and calculates the time required to recover the initial project investment out of the subsequent cash flow. It is computed by dividing the investment amount by the sum of the annual returns (income – expenditure) until it is equal to the capital cost.

Net Present Value (NPV): Net Present Value is considered as one of the important measure for deciding the financial viability of a project. The net present value (NPV) of the project is the difference between two values i.e. discounted cost- discounted returns. Higher the value of NPV is always desirable for a project.

Benefit-Cost Ratio (B-C Ratio) or Profitability Index (PI): The B-C Ratio also referred as Profitability Index (PI), reflect the profitability of a project and computed as the ratio of total present value of the returns to the total present value of the investments (B/C). Higher the ratio better is the return.

Internal Rate of Return (IRR): A measure for deciding the financial viability of a project or an investment. Internal Rate of Return (IRR) indicates the limit or the rate of discount at which the project total present value of return equals to total present value of investments i.e. (project total present value of return

- total present value of investments) = Zero. In other words it is the discount rate at which the NPV of the project is zero. The IRR is computed by iteration i.e. computing NPV at different discount rate till the value is nearly zero. It is desirable to have projects with higher IRR.

Risk and Uncertainty: Risk and Uncertainty are associated with every project. Risk is related to occurrence of adverse consequences and is quantifiable. It is analysed through probability of occurrences. Whereas uncertainty refers to inherently unpredictable dimensions and is assessed through sensitivity analysis. It is therefore necessary to analyse these dimensions during formulation and appraisal phase of the programme. Factors attributing to risk and uncertainties of a project are grouped under the following;

- **Technical-** Relates to project scope, change in technology, quality and quantity of inputs, activity times, estimation errors etc.
- **Economical-** Pertains to market, cost, competitive environment, change in policy, exchange rate etc.
- **Socio-political-** Includes dimensions such as labour, stakeholders etc.
- **Environmental factors:** Could be level of pollution, environmental degradation etc.

Project Appraisal: The project appraisal is the process of critical examination and analysis of the proposal in totality. The appraisal goes beyond the analysis presented in the feasibility report.

Project Management Techniques: It can be classified under two broad categories i.e., Bar Charts and Networks.

Bar Charts: Bar charts are the pictorial representation of various tasks required to be performed for accomplishment of the project objectives.

Gantt Chart: Henry L Gantt around 1917 developed a system of bar charts for scheduling and reporting progress of a project. It is a pictorial representation specifying the start and finish time for various tasks to be performed in a project on a horizontal timescale. Each project is broken down to physically identifiable and controllable units, called the Tasks. These tasks are indicated by means of a bar, preferably at equi-distance in the vertical axis and time is plotted in the horizontal axis.

Milestone Chart: The milestone, represented by a circle over a task in the bar chart indicates completion of a specific phase of the task. In a milestone chart a task is broken down in to specific phases (activities) and after accomplishment of each of the specific activity a milestone is reached or in other words an event occurs. The chart also shows the sequential relationship among the milestones or events within the same task but not the relationship among milestones contained in different tasks.

Networks Analysis: The network is a logical extension of Gantt's milestone chart to illustrate interrelationship between and among all the milestones in an entire project. The two techniques for network analysis are Programme Evaluation and Review Technique (PERT) and Critical Path Method (CPM). These two techniques were developed almost simultaneously during 1956-1958.

PERT (probabilistic) was developed for US navy for scheduling the research and development activities for Polaris missiles programme. PERT is used where emphasis is on scheduling and monitoring the project.

CPM (deterministic) was developed by E.I. du Pont de Nemours & Company as an application to construction project. CPM is used where emphasis is on optimizing resource allocation.

Both CPM and PERT describe the work plan of project where arrows and circles respectively indicate the activities and events in the project.

Terminology used in PERT

Activity: A recognizable work item of a project requiring time and resource for its completion.

Dummy Activity: An activity that indicates precedence relationship and requires no time nor resource.

Critical Activity: Activities on the critical path having zero slack / float time.

Critical Path: The longest time path connecting the critical activities in the project network. The total time on this path is the shortest duration of the project.

Event: An instantaneous point in time signifying completion or beginning of an activity.

Burst Event: An event which gives rise to more than one activity.

Merge Event: The event which occurs only when more than one activity are accomplished.

Expected Time: The weighted average of the estimated optimistic, most likely and pessimistic time duration of a project activity:

$$\text{Expected Time}\left(T_E\right)=\frac{To+4Tm+Tp}{6}$$

where, To is the Optimistic time,

T_M is the Most likely time

T_P is the Pessimistic time

Earliest Start Time (EST): The earliest possible time at which the event can occur. The EST also denotes the Earliest Start Time (EST) of an activity as activities emanate from events. **Latest Start Time (LST):** The latest time at which the event can take place. Also referred as the Latest Start Time (LST) indicating the latest time at which an activity can begin without delaying the project completion time.

Slack: The amount of spare time available between completion of an activity and beginning of next activity.

Resource Analysis: By associating with each activity in the network, the units of resources that would normally be used to carry it out, it is possible to analyse the total resources requirement against time over the duration of the project. From a schedule where all activities start at their earliest start times, a resource histogram can be constructed, showing the total units of resources required for each unit of time through the project. This procedure is called resource analysis.

Project Crashing: Project crashing is an exercise carried out to reduce the time of a project by investing more money or other resources. This becomes necessary when the dead line has to be met. For crashing only the critical are considered since duration of the project could be reduced by crashing these activities only.

Monitoring: Monitoring is a management function, which begins with the start of a project and ends with the completion of the project but it is a continuous process during the implementation of project. The key requirement for monitoring is an ACTION PLAN without which monitoring is not possible.

Evaluation: The world evaluation has its origin in the Latin root "Valere" - to be strong or 'valiant'. Mainly performed by outside agency at various level of project execution. Literally, evaluation means 'assessing the value of'. Evaluations are undertaken in all spheres of life, in informal or formal ways. Evaluation provides information about past or ongoing activities as a basis for modifying or redesigning future strategies. It is a learning process.

Levels of evaluation

Ongoing/process evaluation: When the work is in progress.

Outcome evaluation: Evaluation after completion of the work.

Degree of evaluation

1. Casual every day: - Eg: - Test of Good meal
2. Self Checking: - Eg: - Getting other peoples judgments
3. Do it yourself: - Eg: - Survey or score card

Types of Evaluation

Formal	**Informal**
Systematic appraisal of determining the worth, value of some things	Casual, every day evaluation involves mental assessment on observation or facts
Formative	**Summative**
Attempt to assess the Demerits of programme during Implementation stage	Assess the worth of the final result of programme
Ongoing	**Ex post facto**
Action oriented analysis of a project carried out during implementation	Resume the effect several year after completion of the project

Appraisal and Monitoring

Appraisal is usually taken to mean a critical examination of a proposal of a programme or project, normally before the latter is approved for implementation and funding, appraisal deals with economic viability, technical feasibility and social desirability of a project that is being contemplated.

Monitoring and evaluations are undertaken to find out how a programme or project performs or has performed. Monitoring can be broadly defined as frequent largely routine-wise collection and analysis of and reporting on information about the performance of the work in a programme or project. Thus, monitoring aims at meeting information requirements for current programme and project management.

Social Audit: A social audit is an ongoing process by which the potential beneficiaries and other stakeholders of an activity or project are involved from the planning to the monitoring and evaluation of that activity or project.

Indicators: Measures of 'change', progress and achievement in a project or programme.

Microfinance: Providing the small amount of loans that are repaid within short period of time and are essentially used by low income individuals and households.

Management by Objectives (MBO): Also known as **management by results (MBR)**, is a process of defining objectives within an organization so that management and employees agree to the objectives and understand what they need to do in the organization in order to achieve them. The term "management by objectives" was first given by Peter Drucker.

Halo effect: It is a cognitive bias in which an observer's overall impression of a person influences the observer's feelings and thoughts about that person's character. It was named by psychologist Edward Thorndike in reference to a person being perceived as having a halo.

Development administration: It is an action-oriented and goal-oriented administrative system. It was first coined by U.L. Goswamy in 1955, but the formal recognition to it was given when the Comparative Administration Group of the American Society for Public Administration and the Committee on Comparative Politics of the Social Science Research Council of the USA laid its foundation.

According to Edward Weidner defined "the process of guiding an organization towards the achievement of progressive political, economic and social objectives that are authoritatively determined in one manner or the other."

As far as development administration of India is concerned, system of governance apart, is of concern to people is an improvement in living standards and amenities like water supply, power, a reliable means of transport, a good education system that provides real education to all children and college students, health facilities that are easily available, and are clean and actually cure people or help them recover from illness and injury.

Transactional Analysis (TA): It is an integrative approach to the theory of psychology and psychotherapy. It is described as integrative because it has elements of psychoanalytic, humanist and cognitive approaches. TA was first developed by psychiatrist Eric Berne, in the late 1950s.

According to the International Transactional Analysis Association, TA 'is a theory of personality and a systematic psychotherapy for personal growth and personal change'.

1. As a theory of personality, TA describes how people are structured psychologically. It uses what is perhaps its best known model, the ego-state (Parent-Adult-Child) model, to do this. The same model helps explain how people function and express their personality in their behavior
2. As Berne set his Psychology up, there are four life positions that a person can hold which are:
   a) I'm OK and you are OK: This is the healthiest position about life and it means that i feel good about myself and that i feel good about others and their competence.

b) I'm OK and you are not OK: In this position i feel good about myself but i see others as damaged or less than and it is usually not healthy,

c) I'm not OK and you are OK: In this position the person sees him/ herself as the weak partner in relationships as the others in life are definitely better than the self. The person who holds this position will unconsciously accept abuse as OK.

d) I'm not OK and you are not OK: This is the worst position to be in as it means that i believe that i am in a terrible state and the rest of the world is as bad. Consequently there is no hope for any ultimate supports.

Organizational Climate (Corporate Climate): It is the process of quantifying the "culture" of an organization. It is a set of properties of the work environment, perceived directly or indirectly by the employees, that is assumed to be a major force in influencing employee behavior.

Work Motivation: Work motivation "is a set of energetic forces that originate both within as well as beyond an individual's being, to initiate work-related behavior, and to determine its form, direction, intensity, and duration"

Logical Framework Approach (LFA): It is a management tool mainly used for designing, monitoring and evaluating international development projects. It is also widely known as Goal Oriented Project Planning (GOPP) or Objectives Oriented Project Planning (OOPP).

Job Rotation (Cross training): Job rotation can be defined as lateral transfer of employees among a number of different positions and tasks within jobs where each requires different skills and responsibilities. Individuals learn several different skills and perform each task for a specified time period. Hence, job rotation permits individuals to gain experience in various phases of the business and, thus, broaden their perspective.

Job satisfaction: An individual's general attitude toward his or her job. According to Locke (1976), job satisfaction is a pleasurable or positive emotional state resulting from the appraisal of one's job and job experiences. To measure job satisfaction two indexes were developed i.e.

1. Job Descriptive Index (JDI) (Smith, Kendall, & Hulin, 1969)
2. Minnesota Satisfaction Questionnaire (MSQ) (Weiss, Dawis, England, & Lofquist, 1967).

They measure five important factors conducive to job satisfaction i.e. mentally challenging work, equitable rewards, opportunities for promotion, supportive working conditions, and supportive colleagues.

Total Quality Management (TQM): Consists of organization-wide efforts to install and make permanent climate in which an organization continuously improves its ability to deliver high-quality products and services to customers.

Participatory Approaches in Extension Programme Planning

RRA Methods: The technique used when organizations need local conditions to be assessed as quickly and efficiently as possible. RRA is also appropriate when the aim is to gain an initial orientation, to analyze a specific problem, or to focus on one issue among many. As an alternative to conventional approaches it still fits into "top-down" management schemes. RRA approaches can also be used on a regional aggregation level for situational analysis.

PRA Concentrates entirely on the interests of local communities and on strengthening their decision-making powers. PRA is "bottom-up" approach, and the best way to implement it is within the scope of a community-oriented participatory project or program.

Principle of PRA

1. Participation
2. Flexibility
3. Teamwork
4. Optimal ignorance
5. Systematic
6. Triangulation

Method of Conducting PRA

1. Village transect
2. Agro ecosystem map
3. Agro logical zoning
4. Resource map
5. Seasonal calendar
6. Gender disaggregated seasonal calendar
7. Seasonal analyses
8. Social maps
9. Time line
10. Time trend

11. Mobility map
12. Venn diagram
13. Wealth ranking
14. Stakeholder analyses

Social Map: It focuses on habitation patterns, nature of housing, social infrastructure etc.

Resource Map: It focuses on natural resources in the locality and depicts land, hills, rivers, fields, vegetation etc.

Transect: It provides a cross sectional representation of the different agro ecological zones and their comparison against certain parameters including topography, land type and usage, ownership, problems, opportunities and solutions.

Mobility Map: Explore the movement pattern of an individual, a group or the community. The focus is on where people go and for what?

Time Line: It is used to explore the temporal dimensions from a historical perspective.

Trend Analysis: This method use to explore temporal with focus on change. It captures changes and trends related to certain variables over different spans of time.

Seasonal Diagram: It has been used for temporal analysis a cross annual cycles, with months or seasons as the basic unit of analysis.

Cause and Effect Diagram: It focuses on the causal factors of phenomenon, activity, problem and the effects thereof.

Well-being Ranking Method: Commonly used for ranking and grouping household and communities on the basis of income, wealth, and other perceivable well-being criteria.

Venn Diagram: It is used to study institutional and individual relationship.

Managerial Grid: The **managerial grid model** (1964) is a situational leadership model developed by Robert R. Blake and Jane Mouton. This model originally identified five different leadership styles based on the concern for people and the concern for production. The optimal leadership style in this model is based on Theory Y.

Hawthorne Experiment: Hawthorne Studies by Elton Mayo's in the late 1920s and early 1930s showed the importance of groups in affecting the behavior of individuals at work. Mayo's team conducted a number of experiments involving six female workers. The combinations of results during and after the experiment (ie the increase in the workers' productivity when they were returned to their original working conditions) conclude that workers were motivated by psychological conditions more than physical working condition.

Multiple Choice Questions

1. The word "Management" was derived from root:
   a) Latin b) **French**
   c) Greek d) Sanskrit
2. Literally the word 'Management' is:
   a) **House keeping** b) Personal administration
   c) Store keeping d) Staff placement
3. POSCORB term was developed by:
   a) **Gullick and Urwick** b) Urwick
   c) L. Gulick d) H. Fayol
4. The term 'POSCORB' indicates the:
   a) The activities of government
   b) The principles of administration
   c) **The activities of an administrator**
   d) The objectives of organisation
5. Which law says that work tends to expand if excess time available:
   a) **Parkinson's** b) Marconi
   c) Management by Objective d) Commitment principle
6. For people working at top position which management skill was most needed:
   a) Technical skills b) **Conceptual skills**
   c) Clerical skills d) Leadership skills
7. Management by objective was first put forward by:
   a) Edward Weidner b) Montgomery
   c) **Peter E Druker** d) F W Riggs
8. The business strategy follow by company for liquidate or retrench is called as:
   a) 'SO' strategy b) 'WO' strategy
   c) 'ST' strategy d) **'WT' strategy**

9. The Planning research and action institute is located at:
   a) New Delhi b) Hyderabad
   c) **Lucknow** d) Bhopal
10. The core of planning is:
   a) Span of control b) Coordination
   c) Hierarchy d) **Decision making**
11. A person making prompt and good decisions then the ability is knows as:
   a) Decisiveness b) Initiative
   c) Cohesiveness d) **Social insight**
12. The non programmed decision and programmed decision distinguishes by:
   a) Routine nature b) **Subjectivity**
   c) Structured d) Precedence
13. For the purpose of administration, activities are grouped in to various units this indicates:
   a) Span of control b) Hierarchy
   c) Unity of command d) **Departmentation**
14. When an organizations working in various shifts is indicates the departmentation based on:
   a) Territory b) Function
   c) Product d) **Time**
15. Among which of the following span of control is depends on:
   a) Personality of the supervisors b) Age of the organization
   c) Element of space d) **All of the above**
16. The basic character of informal organization is:
   a) Emotional b) Customary
   c) Personal d) **All of the above**
17. The Zone of indifferent is related with:
   a) Decision making b) Leadership
   c) **Authority** d) Motivation
18. To Take decision and acting upon them are the main function of:
   a) **Line authority** b) Staff authority
   c) Super authority d) Functional authority

19. Unity of command is compliment the principle of:
    a) Supervision
    b) **Scalar chain**
    c) Span of control
    d) Administration

20. Staffing include the:
    a) Recruitment
    b) Promotion
    c) Separation
    d) **All of the above**

21. In the process of recruitment last stage is:
    a) Appointment
    b) **Orientation**
    c) Placement
    d) Promotion

22. With the help of interview, assess the candidate's:
    a) Intellectual capacity
    b) General knowledge
    c) **Personal qualities**
    d) Power of expression

23. With the help of time study, determine the:
    a) Wastage of time
    b) Workers punctuality
    c) F**air day's work**
    d) Give timely assistance

24. The process by which to identify skills or knowledge required by the workers in a organization to perform work for succeed in a job:
    a) **Job analysis**
    b) Item analysis
    c) Factor analysis
    d) None of the above

25. An expression of dissatisfaction with any aspect of organizations is called as:
    a) Stress
    b) Conflict
    c) **Grievance**
    d) Competition

26. Formulation of cliques in an organization leads to:
    a) **Damaging effect**
    b) Performing effect
    c) Stabilizing effect
    d) No effect

27. In an organization one person must receive his or her instruction for a particular operation from only one superior is knows as:
    a) Span of control
    b) **Unity of command**
    c) Hierarchy
    d) Power

28. The staff agencies perform the functions such as:
    a) Planning
    b) Consultation
    c) Achieving goals
    d) **Both a & b**

29. In an organization, principles that is directly concerned with preventing overlap, conflict and friction is:
    a) Direction
    b) Supervision
    c) **Coordination**
    d) Span of control
30. Brainstorming is a:
    a) **Creative technique**
    b) Motivation technique
    c) Evaluation technique
    d) Reasoning technique
31. Communication for public relations is called:
    a) Across communication
    b) Interpersonal communication
    c) **External communication**
    d) Reasoning communication
32. A mandatory communication is follow the:
    a) Vertical communication
    b) Horizontal communication
    c) **Both a and b**
    d) None of the above
33. Communication provides instructions, information and clarification in an organization is known as:
    a) Informal communication
    b) Formal communication
    c) **Administrative communication**
    d) Both a and b
34. The term "entrepreneur" is given by:
    a) **Richard Cantinlon**
    b) Tony pears
    c) Curt Lewin
    d) Gilford Pinchot
35. People considering entrepreneur when he take:
    a) High risk
    b) Low risk
    c) No risk
    d) **Moderate risk**
36. Where the 'National institute of Entrepreneurship and small business development' is located:
    a) **New Delhi**
    b) Jaipur
    c) Chennai
    d) Mumbai
37. An entrepreneur works within the organizational is known as:
    a) Entrepreneur
    b) **Intrapreneur**
    c) Intrepreneur
    d) Entrapreneur

38. The path-goal approach of leadership given by:

    a) **Robert House** b) Fred E Fiedler
    c) B F Skinner d) L W Porter

39. Among them which leadership styles indicates the 'take the odd man out':

    a) Country club leadership b) Middle road leadership
    c) **Lassez-faire leadership** d) Team leadership

40. According to which leadership theory leaders are born and not made:

    a) Behaviour theory of leadership
    b) **Great man theory of leadership**
    c) Situational theory of leadership
    d) None of the above

41. Contingency model of leadership was given by:

    a) Rober House b) **Fred E Fiedler**
    c) B F Skinner d) L W Porter

42. Which theory state that leadership is multidimensional:

    a) Behavioural theory b) **Situational theory**
    c) Trait theory d) None of the above

43. In Linkert's four systems of management, 'Benevolent autocraft' indicates:

    a) System 1 b) **System 2**
    c) System 3 d) System 4

44. Who given the theory of 'Transformational leadership':

    a) Robert House b) **Bass, B. M.**
    c) B F Skinner d) L W Porter

45. The term 'motivation' indicates:

    a) **To move or incite into action** b) To bring forward
    c) To energize inner self d) To act according to wishes

46. Term includes under motivation are:

    a) Drives and desires b) Desires and needs
    c) Emotions and feelings d) **Both a and b**

47. Which statement is true regarding Maslow's theory:
    a) **Based on psychoanalysis** b) First theory of motivation
    c) Published in 1942 d) Based on prepotency of needs
48. The process theory of motivation does include the models of:
    a) Victor Vroom b) Herald Kelly
    c) Porter Lawler d) **All**
49. In Maslow's need hierarchy theory, middle level need is:
    a) **Social needs** b) Security needs
    c) Esteem needs d) Self actualization needs
50. In a social system people wants higher need for power, prestige, status and self-confidence indicates the:
    a) **Esteem needs** b) Security needs
    c) Social needs d) Self actualization needs
51. Motivators in Herzberg's need classification indicate the which level of Maslow's need hierarchy:
    a) Social needs b) Security needs
    c) Physiological needs d) **Self actualization needs**
52. Who given the 'Two factor theory' of motivation:
    a) **F Herzberg** b) V H Vroom
    c) J S Adams d) L W Porter
53. The satisfiers according to Herzberg does include:
    a) Achievement b) Recognition
    c) Challenging work d) **All**
54. According to Maslow:
    a) Man perpetually wants animals
    b) A satisfied need does not motivate
    c) **Both a and b**
    d) None
55. According to Herzberg theory, which of the following act as a motivator?
    a) Working conditions b) **Recognition**
    c) Interpersonal relations d) Both a and c

56. Which of the following is latest theory of motivation:
   a) Expectancy theory
   b) **ERG theory**
   c) Theory X and Y
   d) Maturity immaturity theory

57. Scientist who give the 'Valance expectancy theory' of motivation:
   a) C Argyris
   b) **V H Vroom**
   c) A H Maslow
   d) Mc Gregor

58. ERG theory of motivation was given by:
   a) F Herzberg
   b) V H Vroom
   c) **Clayton Alderfer**
   d) L W Porter

59. In the classical theory of motivation, first major exponent is:
   a) Maslow
   b) Fayol
   c) **Taylor**
   d) Mc Gregor

60. The correct statements regarding Herzberg's hygiene factors are:
   a) They are external to the job
   b) They reflect the Adam nature of man
   c) They are related to the job content
   d) **Both a and b**

61. Among them who proposed the need theory of motivation:
   a) David Mc Clelland
   b) **A H Maslow**
   c) Douglas Mc Gregor
   d) None of the above

62. The term 'Management grid and its concept' was given by:
   a) Samuel David
   b) B F Skinner
   c) **Blake and Mouton**
   d) Peter and Lawter

63. Motivation approach 'Carrot and stick' can be linked with;
   a) Natural learning methods
   b) **Behavior control methods**
   c) Controlled exposure methods
   d) Expository methods

64. Among them who proposed the 'Reinforcement theory of motivation':
   a) F Herzberg
   b) V H Vroom
   c) **B F Skinner**
   d) L W Porter

65. The term 'Theory X' was given by:
   a) **Mc Gregor**
   b) Marks
   c) Abrams
   d) L W Porter

66. The assumption of ‘Workers must be guided into being productive’ is from:
    a) **Theory X** b) Theory Y
    c) Theory Z d) Balance theory
67. The central principle of Mc Gregor’s theory Y is:
    a) Direction of control b) Coordination
    c) Scalar principle d) **Integration**
68. Administration pertaining to:
    a) **Policy making** b) Execution of policy
    c) Giving direction to things d) None of the above
69. Among them who given the term “Developmental administration”:
    a) Edward Weidner b) Montgomery
    c) **George Gant** d) F W Riggs
70. The gangplank idea is given by:
    a) **Fayol** b) Follett
    c) Barnard d) Simon
71. The statement is not true regarding theory-X:
    a) **Man has inherent liking of work**
    b) Man wishes to avoid responsibility
    c) Man must be forced for working
    d) Man threatened with punishment
72. Classical theory of organization was developed during:
    a) **1900-1950** b) 1800-1850
    c) 1700-1750 d) None of the above
73. The classical theory of organization given by:
    a) Luther Gulick b) **Henry Fayol**
    c) F W Taylor d) Fayol
74. Among them who conduct the ‘Hawthorne’ experiment:
    a) **Elton Mayo** b) Hugo
    c) F W Taylor d) Fayol
75. According to Marx, class membership was describe by:
    a) Family b) Relationship of the state
    c) Education d) **Relation to the means of production**

76. Classical theory characterized by:
   a) Specialization b) Impersonality
   c) Hierarchy d) **All of the above**

77. The idea of scientific management was given by:
   a) Gilbreth b) Mayo
   c) Lewin d) **Taylor**

78. Scientific management included:
   a) Time and motion studies b) Task analysis
   c) Job Specialization d) **All of the above**

79. Where the Indian Institute of Public administration is located?
   a) **New Delhi** b) Bombay
   c) Bangalore d) Kanpoor

80. Theory X, is concerned with improving:
   a) Output b) Productivity
   c) **Work Efficiency** d) Work environment

81. According to Max Webber, bureaucracy is superior to any other form of organization in:
   a) Precision b) Stability
   c) Discipline d) **All of the above**

82. Among them which theory of management shows systems approach?
   a) Contingency theory b) Integrated theory
   c) Ecological theory d) **Modern theory**

83. Training is an:
   a) Product b) **Process**
   c) Activity d) Programme

84. Which of the following is not under training:
   a) Overt process b) **Covert process**
   c) Communication process d) Learning process

85. In comparison to training, development process is:
   a) Long term b) **Short term**
   c) Equal term d) None of the above

86. During training, shape of learning curve is:
    a) **S-shape** b) U-shape
    c) V-shape d) C-shape

87. For VLW's training programme should essentially focus on:
    a) Design skill b) Conceptual skill
    c) **Technical skill** d) Leadership skill

88. Extension Education institutes mainly provide training to functionaries working at:
    a) Grass root level b) **Middle level**
    c) Top level d) Both a and b

89. Action research programme is an example of:
    a) Formal training b) Direct training
    c) **Non-Formal training** d) In-service training

90. Freshly recruited ARS trainees got NAARM training which is comes under:
    a) Pre training b) **Induction training**
    c) Orientation training d) Refresher training

91. The summer institute is a example of:
    a) Induction training b) Development training
    c) Orientation training d) **Refresher training**

92. Learning by doing is also known as:
    a) Formal training b) **Informal training**
    c) In-service training d) Post-entry training

93. Policy makers and managers mostly used training model:
    a) **Decision model** b) Expert model
    c) Experimental model d) Objective model

94. Who proposed the cyclic model of training:
    a) **Lynton and Pareek** b) Miller R D
    c) York A d) Reay D G

95. Who proposed the system concept of training:
    a) **York A** b) Miller R D
    c) Reay D G d) Kolb D

96. The motivational phase in training is also called as:
    a) Preparatory phase b) Pre-training phase
    c) **Training phase** d) Post training phase

97. In SATNA need assessment method, first step is:
    a) Job description b) **Problem identification**
    c) People identification d) KAS identification

98. The view indicates that "A need is something whose absence or deficiency proves harmful":
    a) Democratic view b) Autocratic view
    c) **Diagnostic view** d) Discrepancy view

99. The literally means of curriculum is:
    a) **Keep on running** b) Content in training
    c) Theory in training d) Printer material

100. In SMART objective, letter 'S' refers to:
    a) **Simple** b) Specific
    c) Sincere d) Secure

101. The first step in SQ3R method is:
    a) Study b) Select
    c) Speculate d) **Survey**

102. Which of the following is the first stage in ELC cycle?
    a) **Experiencing** b) Publishing
    c) Generalizing d) Applying

103. The fulcrum of ELC cycles is:
    a) Experiencing b) **Processing**
    c) Generalizing d) Applying

104. In experimental learning cycle, last phase is:
    a) **Application** b) Generalization
    c) Publication d) Experience

105. In the seven step model of ELC, first and last step are:
    a) **Climate setting and closure**
    b) Experiencing and closure
    c) Experiencing and application
    d) Objective clarification and closure

106. In experimental approach trainer act as:
    a) **Facilitator** b) Contributor
    c) Controller d) Enabler
107. In Sensitivity training needs a time period of:
    a) 15-20 minutes b) 1-2 hours
    c) 1-2 days d) **5-10 days**
108. With the help of non-verbal communication amount of learning takes place is:
    a) 10 percent b) 20 percent
    c) 30 percent d) **40 percent**
109. Win loose game is used to indicates:
    a) Intra-team collaboration b) **Inter-team collaboration**
    c) Inter-personal collaboration d) Intra-personal collaboration
110. In a training progamme, micro lab exercise conducted at:
    a) **Beginning** b) Middle
    c) End d) Any time
111. In a same group, win-loose game made to play with second time then it will gives:
    a) Better results b) Average results
    c) Best result d) **No result**
112. T-group training is an example of:
    a) Transactional method b) **Sensitivity method**
    c) Experimental method d) Laboratory method
113. Among them co-trainer concept of training is very much related with:
    a) **T group** b) L group
    c) O group d) Q group
114. Which of the following training method is liked most by people with divergent learning style:
    a) **Role play** b) Lecture
    c) Clinical study d) Library review
115. In which training method gives more importance to cognitive domain:
    a) **Lecture** b) Role play
    c) Brainstorming d) Workshop

116. At which stage of brainstorming is free flow of ideas takes place:
   a) **Red stage**
   b) Green stage
   c) Yellow stage
   d) Orange stage
117. At which stage of brainstorming evaluation of ideas is takes place:
   a) Red stage
   b) **Green stage**
   c) Yellow stage
   d) Orange stage
118. Who proposed the brainstorming method:
   a) **Osgood**
   b) Gorden
   c) Andrews
   d) Kolb
119. For large audience, training method best suits:
   a) **Discussion method**
   b) Stimulation method
   c) Sensitivity method
   d) Transactional method
120. Computer based training is an example of:
   a) Expository method
   b) Encounter method
   c) Discovery method
   d) **Individualized method**
121. The empathy based training method is:
   a) **Role play**
   b) Case study
   c) Transition analysis
   d) Sensitivity method
122. Which of the following is included under sensitivity training?
   a) T group
   b) L group
   c) Process work
   d) **All of the above**
123. Sensitivity training is not a:
   a) Self-disclosure method
   b) **Trainer dominated method**
   c) Process work method
   d) Both a and c
124. In which training method number of participants is minimum:
   a) Group discussion
   b) Huddle method
   c) **Buzz session**
   d) Symposium method
125. The example of real life training method is:
   a) **Incidence and Case**
   b. Role play and Case
   c) Incidence and Craftsmanship
   d) Study tours and case
126. In which training method emphasis given on problem diagnosis:
   a) **Case method**
   b) Role play
   c) Group discussion
   d) Project method

127. Among them which training method gets its name from a particular behaviour of honeybee:
   a) Brain storming
   b) Huddle method
   c) **Buzz method**
   d) Role play
128. Example of on-the-job training is:
   a) Craftsmanship
   b) Process work
   c) Apprenticeship
   d) **Both a and c**
129. Syndicate is a:
   a) Telling method
   b) **Doing method**
   c) Showing method
   d) Reading method
130. In an role play method individual plays his own role is called:
   a) Socio-drama
   b) Role play
   c) **Psycho-drama**
   d) Arto-drama
131. Time require for conducting buzz session is:
   a) **5-10 minutes**
   b) 15-20 minutes
   c) 25-30 minutes
   d) 55-60 minutes
132. Computer based training use the:
   a) **Digital image**
   b) Analogue image
   c) Iconic image
   d) Enactive image'
133 In brainstorming session trainer act as:
   a) Stimulator
   b) **Facilitator**
   c) Conceptualizer
   d) Controller
134. In role play method data are:
   a) **Self generated**
   b) Brought from outside
   c) Standardized
   d) General
135. Maximum trainer-trainee ratio is in:
   a) Syndicate
   b) Brainstorming
   c) Case
   d) **Lecture**
136. Minimum trainer-trainee ratio is in:
   a) Syndicate
   b) **Brainstorming**
   c) Case
   d) Lecture
137. Training and visit system organized at:
   a) Weekly basis
   b) **Fortnightly basis**
   c) Monthly basis
   d) Half yearly basis

138. The highest feedback from the participants is in:

a) **Instrumental method** b) Discussion method

c) Simulation method d) Field method

139. Role play methods come under:

a) Projective method b) Simulation method

c) Encounter method d) **All of the above**

140. The basis of 'Lynton and Pareek' classification of training method is on:

a) **Involvement of trainee** b) Cognitive emphasis

c) Sensory capacity d) Involvement of trainer

141. Firstly the concept of microlab was used in:

a) **USA** b) France

c) Germany d) India

142. When persons from the same department selected as trainers for the T-group, then this group is:

a) Cousin group b) Father group

c) **Family group** d) Stranger group

143. When only social scientists are selected as trainers in the T-group, then this group is:

a) **Cousin group** b) Father group

c) Family group d) Stranger group

144. The training methods classified of by Pareek and Rao is based on:

a) Cognitive emphasis b) Learner activity

c) Use of experience d) **All**

145. The training method 'Hypnosis' is classified under:

a) Expository method b) Individual method

c) Controlled exposure method d) **Behaviour control method**

146. Among them which not included under entrepreneurship development training phase:

a) Stimulus phase b) **Supplementary phase**

c) Sustenance phase d) Support phase

147. Which training methods suitable for large group:

a) **Buzz session** b) Brain storming

c) Debate d) Symposium

148. Among them which life position shows over dominance?
   a) I am OK you are OK b) **I am OK you are not OK**
   c) I and you are not OK d) I am not OK but you are OK
149. When panel discussion is modified in which three or more resource persons discuss a specific topic is called:
   a) Symposium b) **Colloquium**
   c) Workshop d) Brainstorming
150. In an area, special qualified peoples pool their experience and an opinion among group is called:
   a) Symposium b) **Conference**
   c) Workshop d) Brainstorming
151. When trainer communicates to a group of participants through the spoken words with audio-visual aids and discussion is:
   a) Group discussion b) **Lecture**
   c) Brainstorming d) Workshop
152. A topic presented by speaker to a group of audience verbally is called:
   a) **Lecture** b) Seminar
   c) Panel discussion d) Workshop
153. When large group divided into smaller, in which a topic is discussed within a limited period of time:
   a) Colloquium b) Seminar
   c) **Buzz session** d) Workshop
154. Who was given the objective – training evaluation model:
   a) **Ralph W Tyler** b) Daniel Stufflebeam
   c) Robert Stake d) Michael Scriven
155. Training Performance Index does include:
   a) Training Effectiveness Index b) Training Utility index
   c) Training participation index d) **All**
156. Measuring performance against predetermined goals is called:
   a) Management b) **Evaluation**
   c) Measurement d) Supervision
157. The term evaluation has its root:
   a) **Latin** b) Greek
   c) Persian d) Roman

158. The pilot study in extension programme indicates the:
   a) Post evolution
   b) Concurrent evolution
   c) **Pre evaluation**
   d) Monitoring
159. In the process of programme planning, last stage is:
   a) Evaluation
   b) Auditing
   c) Supervision
   d) **Reconsideration**
160. The official examination of records is called:
   a) Evaluation
   b) Supervision
   c) **Audit**
   d) Appraisal
161. In order to accomplish particular objective, specification of work to be done:
   a) Plan
   b) **Project**
   c) Programme
   d) Objective
162. The father of project management:
   a) **Henry Gantt**
   b) Thomas Pele
   c) Paul Leagans
   d) Booz Allen Hamilton
163. In project management, element that take time, effort and resource to move from one event to another is:
   a) Event
   b) **Activity**
   c) Critical Path
   d) Slack
164. A point which shows the start or completion of one or more tasks is:
   a) **Event**
   b) Activity
   c) Critical Path
   d) Slack
165. Assuming everything goes well then the largest possible time to perform an activity is:
   a) Optimistic Time
   b) **Pessimistic Time**
   c) Most likely time
   d) Expected time
166. From the initial event to the terminal event, longest possible continuous pathway is known as:
   a) Event
   b) Activity
   c) **Critical Path**
   d) Slack
167. The average time would require by the task if the task were repeated on a number of occasions over an extended period of time:
   a) Optimistic time
   b) Pessimistic time
   c) Most likely time
   d) **Expected time**

168. An activity which shows only relations among the activities is:
   a) Predecessor activity
   b) Successor activity
   c) **Dummy activity**
   d) Nominal activity
169. The basic form of programme evaluation and reviewing technique (PERT) is:
   a) Opportunistic
   b) **Probabilistic**
   c) Deterministic
   d) All of the above
170. The orientation of critical path method (CPM) is:
   a) Opportunistic
   b) Probabilistic
   c) **Deterministic**
   d) All of the above
171. Triangle of project management does include:
   a) Quality
   b) Time
   c) Cost
   d) **All**
172. In project management, gantt charts are used for:
   a) Project Evaluation
   b) Input utilization
   c) **Project scheduling**
   d) Policy formulation
173. The logical framework technique is used in:
   a) **Project planning**
   b) Input utilization
   c) Work scheduling
   d) Policy formulation
174. Among them who give the concept of PRA:
   a) **Henry ford**
   b) Robert chambers
   c) Neils Roling
   d) Nila Mukerjee
175. The father of PRA:
   a) Neils Rolling
   b) **Robert Chambers**
   c) Van den Ban
   d) Paul Leagans
176. Where the term PRA was first used:
   a) **Kenya**
   b) USA
   c) UK
   d) Israel
177. PRA originate in late:
   a) Sixties
   b) Seventies
   c) **Eighties**
   d) Nineties

178. Among them who introduced the PRA in India:
   a) Government departments
   b) ICAR
   c) **NGOs**
   d) Ford foundation

179. The innovation follow in PRA is:
   a) Methods
   b) **Behaviour**
   c) Process
   d) Actors

180. The innovation follow in RRA is:
   a) **Methods**
   b) Behaviour
   c) Process
   d) Actors

181. In PRA extension personnel's act as:
   a) Elective
   b) Extractive
   c) **Participatory**
   d) None of the above

182. Among them which of the following is not the characteristic of PRA:
   a) Quick
   b) **Rigid**
   c) Interactive
   d) Collective

183. Data collection in RRA is done by:
   a) Extension workers
   b) **Multi disciplinary team**
   c) Research scientists'
   d) Progressive farmers

184. The principle of PRA is:
   a) Optimal ignorance
   b) Triangulation
   c) Offsetting bias
   d) **All**

185. In PRA, cause effect relationship shown by:
   a) Venn diagrams
   b) **Flow diagram**
   c) Social mapping
   d) Livelihood analysis

186. The social calendar is a tool used in:
   a) **PRA**
   b) RRA
   c) LFA
   d) FSR

187. The degree of importance of individuals and institutions in a area shown by which PRA technique:
   a) Social mapping
   b) **Venn diagram**
   c) Semi structured interview
   d) None of the above

188. In a venn diagram circles shows very strong co-ordination will have:
    a) Separate circles
    b) Touching circles
    c) Small overlap
    d) **Larger overlap**

189. The author of book "Participatory Rural Appraisal-Methodology and Applications" is:
    a) **Neela Mukherjee**
    b) Anitha Jamthani
    c) Y P Singh
    d) Baldeo Singh

190. The micro-unit of an agro-ecosystem is:
    a) District
    b) **Village**
    c) Farmer
    d) Household

191. ________ can be used to analyze the competitive situation?
    a) Johari window
    b) TOWS matrix
    c) SWOT analysis
    d) All the above

192. Who propounded 'Borderless flat world'?
    a) Stoner
    b) Thomas Friedman
    c) Vilfredo Pareto
    d) Peter Drucker

193. 'Making known to outside agencies unethical company practices' is generally known as _______
    a) Malfunctioning
    b) Horn-blowing
    c) Whistle-blowing
    d) Ear-blowing

194. 'Chaebol' is characterized by a tight collusion between whom?
    a) **Government and industrial conglomerates**
    b) People and industrial conglomerates
    c) NGO and industrial conglomerates
    d) Government and foreign industrialists

195. In portfolio matrix weak competitive position and high business growth rate is denoted by –
    a) Cash cows
    b) Question mark
    c) Dogs
    d) Stars

196. The selection of a course of action from among alternatives is called as _______
    a) Decision Making
    b) Controlling
    c) Planning
    d) Performing

197. Cost effectiveness analysis tries to achieve which of the following?

a) The best ratio of benefit and cost b) Highest profit

c) Lowest cost d) Zero cost

198. Is an individual who transforms a dream or an idea into a prorfitable venture by employing innovation and creativity.

a) Entrepreneur b) Interpruneur

c) Tycoon d) Manager

199. Who was considered as first lady of management?

a) Elizabeth Taulor b) Indira Noyee

c) Lillian Gilbreth d) Margaret Thatcher

200. Clan control is based on which of the following variables?

a) Norms b) Shared values

c) Expected behavior **d) All of these**

201. Which chart shows the time relationships between the events of a production program?

a) Bar Chart b) Pie Chart

c) Flow Chart **d) Gantt Chart**

202. According to Henry Fayol management consists of how many functions?

a) Three b) Four

c) Five **d) Six**

203. The famous book ‘ The wealth of nations’ written by whom?

a) Amartya b) Karl Marx

c) Adam Smith d) Manmohan Singh

204. Who gave the ‘Five mode model of resolution of conflict?

a) Kenneth W Thomas b) Ralph H Kilmann

c) Both a & b d) Swanson

205. Which institute developed Activity Slack Bar Chart?

a) NAARM, Hyderabad b) NIRD, Hyderabad

c) IIM, Ahmadabad d) IIT, Mumbai

206. Costs for controlling environmental language are technically known as

a) Abatement cost b) Reductionist cost

c) Environmental d) None of the above

207. Design skill of manager refers to
    a) Solving problems that benefit the enterprise
    b) Understanding the relationship among the elements
    c) Possessing technical knowledge
    d) Maintaining human relationship with all
208. Which one of the following is a Management Pad?
    a) Quality circle
    b) Demassing
    c) Strategic alliance
    d) All of these
209. When a economy face a rapid unstable growth the situation can be termed as
    a) Boom economy
    b) Bull run
    c) Bubble economy
    d) None of these
210. Behaviour Anchored Rating scale (BARS) is related to
    a) Performance appraisal
    b) Sociometry
    c) Psychological
    d) None of these
211. Effectiveness is measured in
    a) Problem solving ability
    b) Doing things right
    c) Producing results
    d) All of these
212. Technique to forecast possible outcomes for decision making.
    a) Puerto Rico Simulation
    b) Gibraltar Simulation
    c) San Diego simulation
    d) Monte Carlo simulation
213. The 360° performance appraisal was first developed by
    a) Wipro
    b) General Electric (USA)
    c) TATA motors
    d) Hindustan Unilever

9

Information Communication Technology (Theory)

ICT or Information and Communications Technology: The basket of technologies, which assist or support in storage, processing and dissemination/ communication of data/information. ICT thus includes technologies such as desktop and laptop computers, software and connection to the Internet that are intended to fulfil information processing and communication functions.

ICT Tools

- Word processing applications - used for preparing text documents.
- Presentation software - used for preparing presentations
- Spreadsheets - used for calculation and preparing tabular data with calculations
- Databases - used to store the information in the form of records and retrieving information using query mechanism.
- Multimedia – used to organize the information more meaningful manner with the help of text, images, animation, audio and videos
- Web browsers and E-mail – used to searching information and sending or receiving mail using Internet.

Virtual Community: A virtual community/e-community/online community is a group of people that primarily interact via communication media such as letters, telephone, email or net rather than face to face. If the mechanism is a computer network, it is called an online community.

Virtual Extension Research Communication Network (VERCON): Virtual Extension and Research Communication Network (VERCON) established by FAO aims to use the potential of the Internet and apply it to strengthening and enabling linkages between and within agricultural research and extension institutions through human and technological components.

Information Kiosk: The concept of 'ICT kiosk' is to provide various services including agricultural information to the rural people to enhance the accessibility of information at village level.

Computers: The term "Computer" is derived from the word "Compute". The computer can be defined as "an electronic data processing machine that accepts data, process it according to the provided set of instructions and produces the desired output".

Types of Computers

Computers are classified into four categories on the basis of its size viz. Super computers, main frame computers, mini computers, personal or laptop computers.

Super Computers

PARAM by CDAC Pune and SX2 by Nippon are the examples of super computers.

Main frame computers

VAX 8000 series, IBM 4300 series are the examples of main frame computers.

Mini computers

Horizon, PDI 11/70 is the examples of mini computers.

Personal computers and laptop computers

Pentium IV computers with 3GHz and higher speed are available today. Laptops are portable computers which are meant for use by people who need computing power mobile.

Expert system: It is a software application that attempts to reproduce the performance of one or more human experts. Expert systems are mostly based on a specific problem domain, and are a traditional application of artificial intelligence. The expert system is used to behave like a human expert to solve the problem with the help of pre-set conditions in the software application.

A Geographical Information System (GIS): The geographical information system is also called as a geographic information system or geospatial information system. It is an information system capable of integrating, storing, editing, analyzing, sharing, and displaying geographically referenced information.

e-Governance: It means the use of Information and Communication Technology (ICT), particularly web based applications, to provide access to and deliver information / service to the public, business, other agencies and governmental entities faster, cheaper, easier and more efficiently.

e-Government: E-Government or Electronic Government is the delivery of more convenient, customer oriented and cost effective public services and sharing of information through electronic media. This deal with the development of online services to the citizen, i.e. government services – such as e-agriculture, e-passport, e-tax, e-transportation, e-health, etc. **Cyber** - The word Cyber means, "relating to Information Technology, the Internet, and virtual reality, the Cyber Space".

Cyber Space: Cyber Space is the imaginary or virtual space of computers connected with each other on networks, across the globe.

Cyber Extension: Cyber Extension can be defined as the "Extension over Cyber Space". In the applied context of agriculture and animal husbandry, cyber extension means "using the power of online computer networks with the help of communication channels to deliver the content in the form of text, graphics, audio and video either passively or interactively to facilitate dissemination of agricultural/animal husbandry technology".

Knowledge Management System (KM System): Refers to a (IT based) system for managing knowledge in organizations for supporting creation, capture, storage and dissemination of information.

World Wide Web: (**WWW** or commonly known as **the web**) It is a system of interlinked hypertext documents accessed via the Internet. On March 12, 1989, Tim Berners-Lee, a British computer scientist and former CERN employee, wrote a proposal the World Wide Web.

Information Village Research Project (IVRP): The Information Village Research Project (IVRP) was set up in 1998 in Pondicherry, by the M.S. Swaminathan Research Foundation (MSSRF), a nonprofit organization. The project was conceptualized as part of a larger initiative called the "bio-village" project that was initiated in 1991. An interdisciplinary dialogue on information technology — "Reaching the Unreached".

ICT initiatives in India

1. AMUL	Gujrat
2. CARD	Hyderabad
3. SEWA, 1998	Ahmedabad
4. TARAHAAT	Jhansi
5. E-chaupal	Ujjain
6. Akshaya, 2001	Kerala
7. Bhoomi	Karnataka Hyderabad

8. FRIENDS, 2000 Thiruvanantjhapuram
9. Warna wired village, 1999 Maharashtra
10. Gramdoot Jaipur
11. Janmitra, 2002 Jhalawar
12. Mahitishakti, 2001 Panchmahal, Gujrat

CARD: Computer-aided Administration of Registration Department

CICs: Community Information Centers

FRIENDS: Fast, Reliable, Instant, Efficient Network for Disbursement of Services

SEWA: Self Employed Women's Association

Latest information communication technology application in India

 e-Village	• The model e-Village Project is currently being implemented in selected 10 remote tribal villages of East Siang District of Arunachal Pradesh, North-East India, from July 2008. • The main aim of this project was creating computer infrastructure in remote tribal villages to provide e-awareness and e-literacy to the tribal community, developing an integrated sustainable model for ICT based services for North Eastern States with special focus on Agriculture and allied areas. • Regular computer training classes are going to be conducted to the children, village youth and others.
	• The proposed touch screen kiosk provides the flexibility in providing information on various mode of farming practices including the crops, commoditi.0es, and enterprises. • The accessibility to the agricultural technical information without disrupting the routine engagement of farmers in their farm activities.
 e – Arik	• The project proposes single window system for the improved agricultural information and technology delivery at the selected village by using computer, internet, phone, radio and television. • Project proposes to provide all time expert consultation on agriculture production, protection and marketing aspects through ICTs.

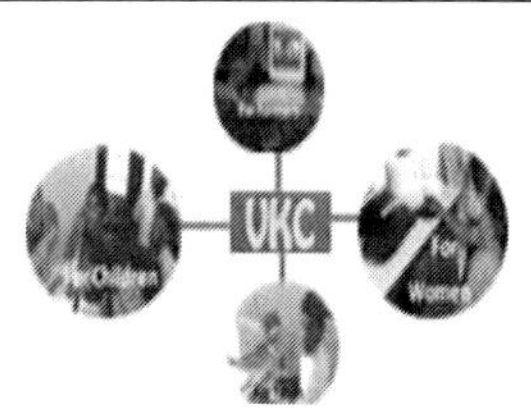 Village Knowledge Canters	• Started in January 1998, Pondicherry. Village Knowledge Centres serve as information dissemination centre providing instant access to farmers to latest information/ knowledge available in the field of agriculture/A.H., starting from production to marketing. • Reaching the unreached.
 Village Resource Canters	• To provide the space technology enabled services directly to the rural population, ISRO launched the Village Resource Canters (VRCs, October 18, 2004) programme in association with reputed NGOs/ Trusts and state/ central agencies. 473 VRCs have been set up in 22 States/Union Territories. VRCs also provide a variety of services like tele-education, telemedicine, online decision support, interactive farmers' advisory services, tele-fishery, e-governance services, weather services and water management.
 	• A call centre based extension service will be delivering knowledge and information exactly as per the requirements of the farming community which from 21.1.04. • The Call Centers can be accessed by farmers all over the country on common Toll Free Number 1551. • Since 10th June, 2004 the Call Centers service has been made available right from 6 A.M. to 10 P.M. except on Sundays and gazetted holidays, beyond these hours the calls are attended in the IVRS mode.
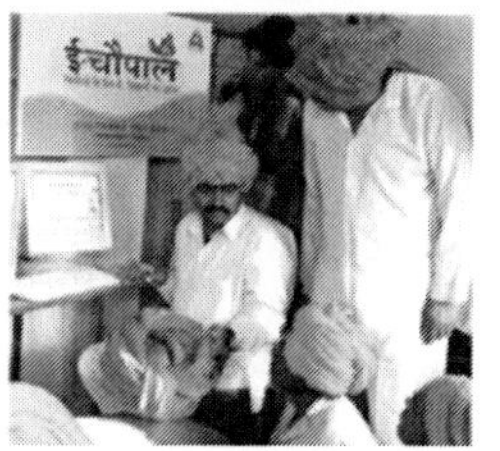	• e-Choupal ' The unique web based initiative of ITC 's Agri Business Division, offers the farmers of India all the information, products and services they need to enhance farm productivity, improve farm gate price realization and cut transaction costs. • The programme installs computers with Internet access in rural areas of India to offer farmers up-to-date marketing and agricultural information. • Launched in June 2000, 'e-Choupal', has already become the largest initiative among all Internet-based interventions in rural India.

Multiple Choice Questions

1. MASS imparting the
   - a) Training
   - b) Incentives
   - c) Demonstration
   - d) **All**
2. Warna wired village are linked through:
   - a) PAN
   - b) **WAN**
   - c) LAN
   - d) None
3. Akshaya started in:
   - a) New Delhi
   - b) Hyderabad
   - c) Lucknow
   - d) **Kerala**
4. Community Information Centers started in:
   - a) Madhya Pradesh
   - b) Chandigarh
   - c) **Sikkim**
   - d) All
5. Bhoomi project started in:
   - a) **Karnatak**a
   - b) Andhra Pradesh
   - c) UP
   - d) All
6. National e-governance plan of India started in:
   - a) 2001
   - b) **2003**
   - c) 2005
   - d) 1999
7. E-seva in Hyderabad also know as:
   - a) FRIENDS
   - b) **TWINS**
   - c) GRAMDOOT
   - d) None
8. Gyandoot started in:
   - a) Ahmedabad
   - b) **Madhya Pradesh**
   - c) Tamilnadu
   - d) None
9. PAN network used among:
   - a) With in cities
   - b) **With in room**
   - c) With in country
   - d) All
10. The first newspaper started in India was:
    - a) **Bengal Gazette**
    - b) Times of India
    - c) Rural India
    - d) Bengal Outlook
11. Agricultural Research Information System Network is:
    - a) HORTNET
    - b) **ARISNET**
    - c) AGRISNET
    - d) VISTARNET

12. Agricultural Extension Information system Network is:
    a) CoopNET b) FERTNET
    c) AGRISNET d) **VISTARNET**
13. The toll free telephone number for kissan call center is:
    a) **1551** b) 1331
    c) 5115 d) 3113
14. Warna Village Network project was started in the state of:
    a) Utter Pradesh b) **Maharashtra**
    c) Tamil Nadu d) Madhya Pradesh
15. The e-chaupal concept in extension delivery is promoted by:
    a) **ITC** b) MSSRF
    c) KINFRA d) CAPART
16. The concept of Selective Dissemination of Information (SDI) was introduced by
    a) Kaiser **b) Luhn of IBM**
    c) S. R. Ranganathan d) Derek Austin.
17. Binary system of numbers uses only
    a) **Two digits** **b)** Four digits
    c) Eight digits d) Six digits
18. Information is …
    a) Raw data b) Raw knowledge
    c) Input data d) **Organized data**
19. To which country the credit is given to coin the term information society?
    a) USA c) France
    b) **Japan** **d)** India
20. What is the unit of information?
    a) **Bit** c) Byte
    d) Gram d) Hertz

21. The invisible web refers to-
    a) The internet, since we cannot see it
    b) **That part of the internet, which is hidden from the search engines**
    c) The telecommunication signals which are not seen
    d) The failure in accessing the web pages
22. The act enacted in India in 1856 on Intellectual Property Right was based on.
    a) American Patent Law 1810
    b) British Patent Law 1852
    c) The Patent Bill
    d) The Design Act of 1911
23. Whether intellectual property can be sold.
    a) No
    c) **Sale is possible**
    b) Yes
    d) None of these
24. The term communication came from which language?
    a) Greek
    c) **Latin**
    b) German
    d) French
25. ASK (Approach, Skill, and Knowledge) principle is related to
    a) Classification
    b) Management
    c) ICT
    c) Standards
26. IC chips used in computers are usually made of
    a) Lead
    b) Silicon
    c) Chromium
    d) Gold
27. One kilobyte is equal to
    a) 1000 bytes
    b) **1024 bytes**
    c) 100 bytes
    d) 1023 bytes
28. Which of the following is not an example of Operating System?
    a) Windows 98
    b) BSD Unix
    c) **Microsoft Office XP**
    d) Red Hat Linux
29. Which supercomputer is developed by the Indian Scientists?
    a) **Param**
    b) Super 301
    c) Compaq Presario
    d) CRAY YMP

30. One Gigabyte is Approximately equal is

a) 1000,000 bytes
b) **1000,000,000 bytes**
c) 1000,000,000,000 bytes
d) None of these

31. Check the odd term out

a) Internet
b) Linux
c) Unix
d) Windows

32. What is the control unit's function in the CPU

a) To transfer data to primary storage
b) TO store program instructions
c) **To perform logic functions**
d) To decode program instructions

33. What is the full form of www?

a) Worldwide word
b) **World wide web**
c) Word winding works
d) Words wide Spiderman wez

34. Which of the following commands is given to reboot the computer?

a) **Ctrl + Alt + Del**
b) Ctrl + Alt + Tab
c) Ctrl + Shift + Del
d) Ctrl + Alt + shift

35. Which of the following statements is true?

a) **Minicomputer works faster than Microcomputer**
b) Microcomputer works faster than Minicomputer
c) Speed of both the computers is the same
d) The speeds of both these computers cannot be compared with the speed of advanced computer

36. The background of any word document -

a) Is always white colour
b) **Is the colour you preset under the Options menu**
c) Is always the same for the entire document
d) Can have any colour you choose

37. The blinking symbol on the computer screen is called the

a) Mouse
b) logo
c) Hand
d) **Cursor**

38. Select the odd one out -
    a) Interpreter
    b) **Operating System**
    c) Compiler
    d) Assembler
39. CRM means -
    a) **Customer Relationship Management**
    b) Customer Retention Manager
    c) Customers' Relatives Meet
    d) Channel Route Market
40. A presentation means
    a) Display of products
    b) **Explaining the utility of products or plan**
    c) A gift
    d) Display of Communication skils
41. The term PC means
    a) Private Computer
    b) Personal Computer
    c) Professional Computer
    d) Personal Calculator
42. Silicon Valley of India is located in
    a) Dehradun
    b) Bangalore
    c) Hyderabad
    d) Srinagar
43. ROM is composed of
    a) Magnetic cores
    b) **Micro-processors**
    c) Photoelectric cells
    d) Floppy disks
44. Which one of the following is not a computer language?
    a) BASIC
    b) COBOL
    c) **LOTUS**
    d) FORTRAN
45. Which one of the following is not a programming language of a computer?
    a) BASIC
    b) FORTRAN
    c) LASER
    d) PASCAL
46. The brain of a computer is
    a) **CPU**
    b) CD
    c) Floppy disc
    d) Megabyte

47. The most widely used commercial programming computer language is
    a) BASIC   **b) COBOL**
    c) FORTRAN   d) PASCAL
48. Who was the inventor of mechanical calculator for adding numbers?
    a) Charles Babbage   b) Peano
    c) Newton   **d) Pascal**
49. First generation computer systems used
    a) Transistors   **b) Vacuum tubes**
    c) Magnetic cores   d) Silicon chips
50. Which is/are the main provisions of Information Technology (IT) Act, 2000?
    a) Finalisation of contracts through electronic media
    b) Legal recognition to digital signatures and documents
    c) Making hacking with computer system an office
    d) All of the above
51. A computer programmer
    a) Enters data into computer
    b) Writes programs
    c) Changes flow chart into instructions
    d) Does total planning and thinking for a computer
52. WAN means -
    a) Wide Area Network   b) Wine Area Network
    c) Woll Area Network   d) Wide Allen Network
53. MAN is a -
    a) Cost   b) Company
    c) **Network**   **d)** Country
54. Which of the following is the universal gate?
    a) NAND-Gate   b) OR-Gate
    c) AND-Gate   d) NOT-Gate
55. What is part of a database that holds only one type of information?
    a) Report   **b) Field**
    c) Record   d) File

56. 'OS' computer abbreviation usually means ?
    a) Order of Significance b) Open Software
    c) Operating System **d)** Optical Sensor
57. Internet in India introduced by
    a) **VSNL** b) BSNL
    c) Both d) None
58. NKC used for:
    a) National Knowledge Commission
    b) National Knowledge Core
    c) National Knowledge Center
    d) None
59. The e-Governance plan runs in the years:
    a) 2000-2004 b) **2003-2007**
    c) 2008-2012 d) 2013-2017
60. DACNET signed on:
    a) 12 September 2004 b) 12 September 2002
    c) **12 September 2001** d) 12 September 2007
61. Twin cities Integrated Network Systems also known as:
    a) GRAMDOOT b) GYANDOOT
    c) **E-SEVA** d) CARD
62. Janmitra project is a joint initiative of:
    a) GOI & UNESCO b) **UNDP & GOI**
    c) UNDP & UNESCO d) All
63. Village Knowledge Centres started with the support of:
    a) IDRC b) CIDA
    c) **Both a & b** d) None
64. Knowledge on wheels launched in a year:
    a) 2004 b) 2011
    c) 2007 d) 2008
65. Marketing Research and Information Network also known as:
    a) **AGMARKNET** b) TNAGMARK
    c) NICNET d) All

66. AGMARKNET launched in the year:
   a) **March, 2000**
   b) July, 2002
   c) June, 2001
   d) April, 2006
67. e-Arik implemented in which village of India:
   a) **Yagrung**
   b) Dhar
   c) Bichiya
   d) None
68. In KISSAN Kerala 'KISSAN' stands for:
   a) Karshaka Information Systems, Services and Networking
   b) Kerala Information Systems, Services and Networking
   c) Karshaka Information Systems, Services and Network
   d) Kerala Information Systems, Services and Network
69. Kisan Mobile Sandesh was started in the year:
   a) **2008**
   b) 2010
   c) 2012
   d) 2006
70. CARD stands for:
   a) Computerized Administration of Registration Department
   b) Computer-aided Administration of Registration Department
   c) Common Administration of Registration Department
   d) Complete Administration of Registration Department
71. CICs full form is:
   a) Community Intrest Centres
   b) Community Information Centres
   c) Common Information Centres
   d) None
72. In FRIENDS 'I' indicate:
   a) Interest
   b) Insurance
   c) **Instant**
   d) None
73. Full form of SEWA is:
   a) Self Employed Women's Association
   b) Self Employed Women's Agency
   c) Self Employed Women's in Action
   d) Self Employed Women's Assembly

74. "AKASHGANGA" has established by:
    a) IVRI
    b) **NDDB**
    c) ICAR
    d) NDRI
75. AKASHGANGA provides total integrated solution for
    a) **Automatic milk collection**
    b) Automatic sample collection
    c) Automatic semen collection
    c) Automatic milk cooling
76. "AKASHGANGA" has been established in the year:
    a) 1990
    b) 1998
    c) **1996**
    d) 1999
77. Radio Frequency Identification (RFID) used for:
    a) **Animal identification**
    b) Animal slaughtering
    c) Both
    d) None
78. 'Telephone lady' concept was introduced in ....
    a) USA
    b) China
    c) Bangladesh
    d) India
79. APARIS – stands for
    a) Andhara-Pradesh Agricultural Research Information System
    b) Asia-Pacific Agricultural Research Information society
    c) Asia-Pacific Agricultural Research Information System
    d) Asia-Pacific Agricultural Reference Information System
80. The official publication wing of the ICAR is
    a) Division of Agricultural Education, New Delhi
    b) The Directorate of Knowledge Management in Agriculture, New Delhi
    c) NCAP, New Delhi
    d) NAARM, Hyderabad
81. When journals provide Open Access to their articles, it is referred as............
    a) Golden Road
    b) Diamond Road
    c) Open Road
    d) Closed Road
82. A blog is a fusion of words ....
    a) "web" and "googel"
    b) "web" and "log"
    c) "WAN" and "LAN"
    d) "glob" and "log"

83. Commercial GIS software includes which of the following?

a) ArcGIS b) GeoMedia

c) Geomatica Maplnfo **d) All of these**

84. The term "remote sensing" was coined by ....

a) Evelyn Pruitt b) Roger Tomlinson

c) Mark Zuckerberg d) Tim Berners-Lee

85. What is the rank of India in terms of internet users in the world?

a) 2nd **b) 3rd**

c) 4th d) 10th

86. What is the rank of India in terms of number of mobile phones in use in the world?

a) 2nd b) 3rd

c) 4th d) 10th

87. What is the full form of CeRA?

a) The Collection of e-Resources in Agriculture

b) The Consortium for e-Resources in Aquaculture

c) The Consortium for e-Resources in Agriculture

d) The Consortium for e-Readers in Agriculture

88. nLogue was developed to connect e very village in a distinct using terrestrial wireless and provide village kiosks by which institute?

a) IIT, Kanpur b) IIT Mumbai

c) IIT, Chennai d) IIT Hyderabad

89. One-piece mobile computer, operated by touch screen is often referred as

a) Capsule **b) Tablet**

c) Pill d) Pad

90. A graphical representation of a real person is known as

a) Cartoon b) Graphiti

c) Characterization **d) Avatar**

91. What is the full form o USB?

a) United Selection Builder **b) Universal Serial Bass**

c) Unifying Selection Board d) User Specific Benefit

92. 'Cyber Dhaba' concept was started by which organization ?

a) ICAR
b) UNESCO
c) IFFCO
d) ITC

93. Green SIM card for farmers started in 2008 is a joint venture of ....

a) IFFCO and AIRTEL
b) DRDO-ICAR
c) ICAR and C-DAC
d) ITC-Apple

94. In the term 'QR Code' QR stands for

a) Quick Response
b) Quality response
c) Quantitative Response
d) Qualitative Response

Fill in the blanks

- E-agric set up in the year ...**2006**
- Janmitra is joint venture of **UNDP** & **GOI**
- Village Knowledge Center started in **1998**
- ISRO & MSSRF jointly established first **Village resource center i**n 2004.
- Kisan kerala launched in **1ST Nov 2003** at **Punchayat** level.
- DACNET project launched by the **Department of Agriculture & Cooperation.**

10

Statistical Tools in Extension Research (Theory)

Statistics: Collection of methods for planning experiments, obtaining data, and then organizing, summarizing, presenting, analyzing, interpreting, and drawing conclusions.

Sir R.A. Fisher: Father of modern statistics.

Population: All subjects possessing a common characteristic that is being studied.

Sample: A subgroup or subset of the population.

Parameter: Characteristic or measure obtained from a population.

Statistic: Characteristic or measure obtained from a sample.

Karl Pearson: Study of correlation and chi-square test of goodness of fit.

W.S.Gosset: Student's t-test for small sample

Francis Galton: Regression analysis.

Descriptive Statistics: Collection, organization, summarization, and presentation of data.

Inferential Statistics: Generalizing from samples to populations using probabilities. Performing hypothesis testing, determining relationships between variables, and making predictions.

Variable: Characteristic or attribute that can assume different values.

Random Variable: A variable whose values are determined by chance.

Qualitative Variables: Variables which assume non-numerical values.

Quantitative Variables: Variables which assume numerical values.

Discrete Variables: Variables which assume a finite or countable number of possible values. Usually obtained by counting.

Continuous Variables: Variables which assume an infinite number of possible values. Usually obtained by measurement.

Nominal Level: Level of measurement which classifies data into mutually exclusive, all inclusive categories in which no order or ranking can be imposed on the data.

Ordinal Level: Level of measurement which classifies data into categories that can be ranked. Differences between the ranks do not exist.

Interval Level: Level of measurement which classifies data that can be ranked and differences are meaningful.

Ratio Level: Level of measurement which classifies data that can be ranked, differences are meaningful, and there is a true zero. True ratios exist between the different units of measure.

Types of sampling:

Random Sampling: Sampling in which the data is collected using chance methods or random numbers.

Systematic Sampling: Sampling in which data is obtained by selecting every *k*th object.

Convenience Sampling: Sampling in which data is which is readily available is used.

Stratified Sampling: Sampling in which the population is divided into groups (called strata) according to some characteristic. Each of these strata is then sampled using one of the other sampling techniques.

Cluster Sampling: Sampling in which the population is divided into groups (usually geographically). Some of these groups are randomly selected, and then all of the elements in those groups are selected.

Raw Data: Data collected in original form.

Frequency: The number of times a certain value or class of values occurs.

Frequency Distribution: The organization of raw data in table form with classes and frequencies.

Categorical Frequency Distribution: A frequency distribution in which the data is only nominal or ordinal.

Ungrouped Frequency Distribution: A frequency distribution of numerical data. The raw data is not grouped.

Grouped Frequency Distribution: A frequency distribution where several numbers are grouped into one class.

Class Limits: Separate one class in a grouped frequency distribution from another. The limits could actually appear in the data and have gaps between the upper limit of one class and the lower limit of the next.

Class Boundaries: Separate one class in a grouped frequency distribution from another. There is no gap between the upper boundary of one class and the lower boundary of the next class. The lower class boundary is found by subtracting 0.5 units from the lower class limit and the upper class boundary is found by adding 0.5 units to the upper class limit.

Class Width: The difference between the upper and lower boundaries of any class. The class width is also the difference between the lower limits of two consecutive classes or the upper limits of two consecutive classes. It is not the difference between the upper and lower limits of the same class.

Measures of Central Tendency

Mean: Sum of all the values divided by the number of values. This can either be a population mean (denoted by μ) or a sample mean (denoted by x bar). The Mean is used in computing other statistics such as the variance and does not exist for open ended grouped frequency distributions.

Median: The data must be ranked (sorted in ascending order) first. The median is the number in the middle. The Median is the centre number and is good for skewed distributions because it is resistant to change.

Mode: The mode is the most frequent data value. There may be no mode if no one value appears more than any other. There may also be two modes (bimodal), three modes (trimodal), or more than three modes (multi-modal). For grouped frequency distributions, the modal class is the class with the largest frequency.

Measures of Dispersion

Range: The range is the simplest measure of variation to find. It is simply the highest value minus the lowest value. Range = Maximum – Minimum. Since the range only uses the largest and smallest values, it is greatly affected by extreme values, therefore not resistant to change.

Mean Deviation: Average of absolute deviation of all the items of data, from their average. It is used as a measure of dispersion where the number of values or quantities is small, otherwise standard deviation is used.

Mean square deviation

Mean square deviation is defined as the average of square deviations of all the items in the data, taken from an arbitrary value.

Variance and Standard deviation

Population Variance: The average of the squares of the distances from the population mean. It is the sum of the squares of the deviations from the mean divided by the population size. The units on the variance are the units of the population squared.

Sample Variance: It is unbiased estimator of a population variance. Instead of dividing by the population size, the sum of the squares of the deviations from the sample mean is divided by one less than the sample size. The units on the variance are the units of the population squared.

Standard deviation: The population standard deviation is the square root of the population variance and the sample standard deviation is the square root of the sample variance. The sample standard deviation is not the unbiased estimator for the population standard deviation. The unit on the standard deviation is the same as the units of the population/sample.

Coefficient of Variation: Standard deviation divided by the mean, expressed as a percentage.

Standard Error: Standard error is defined as the estimate of standard deviation of sample mean. It is a measure of precision of population parameter μ.

Sheppard's correction for variance: The variable value included in a class is assumed to be concentrated at the middle point. Hence there is a chance of error in the value of estimated variance. Sheppard's correction is used to remove this error.

Population vs Sample: The population includes all objects of interest whereas the sample is only a portion of the population. Parameters are associated with populations and statistics with samples. Parameters are usually denoted using 'Greek' letters (mu, sigma) while statistics are usually denoted using 'Roman' letters (x, s).

Correlation coefficient (r): Independent of origin and change of scale, limits ± 1 and has no unit.

Regression analysis (b): Regression has the unit of dependent variable, no limit and independent of origin but not with scale

Cumulative Frequency: The number of values less than the upper class boundary for the current class. This is a running total of the frequencies.

Relative Frequency: The frequency divided by the total frequency. This gives the percent of values falling in that class.

Cumulative Relative Frequency: The running total of the relative frequencies or the cumulative frequency divided by the total frequency.

Histogram: A graph which displays the data by using vertical bars of various heights to represent frequencies. The horizontal axis can be the class boundaries, the class marks, or the class limits.

Frequency Polygon: A line graph. The frequency is placed along the vertical axis and the class midpoints are placed along the horizontal axis. These points are connected with lines.

Ogive: A frequency polygon of the cumulative frequency or the relative cumulative frequency. The graph always starts at zero at the lowest class boundary and will end up at the total frequency (for a cumulative frequency) or 1.00 (for a relative cumulative frequency).

Pareto Chart: A bar graph for qualitative data with the bars arranged according to frequency.

Pie Chart: Graphical depiction of data as slices of a pie. The frequency determines the size of the slice.

Pictograph: A graph that uses pictures to represent data.

Stem and Leaf Plot: A data plot which uses part of the data value as the stem and the rest of the data value as leaf to form groups or classes. This is very useful for sorting data quickly.

Skewed Distribution: The majority of the values lie together on one side with a very few values (the tail) to the other side. In a positively skewed distribution, the tail is to the right and the mean is larger than the median. In a negatively skewed distribution, the tail is to the left and the mean is smaller than the median.

Symmetric Distribution: The data values are evenly distributed on both sides of the mean. In a symmetric distribution, the mean is the median.

Weighted Mean: The mean when each value is multiplied by its weight and summed. This sum is divided by the total of the weights.

Midrange: The mean of the highest and lowest values. (Max + Min) / 2

Chebyshev's Theorem: The proportion of the values that fall within *k* standard deviations of the mean is at least where k > 1. Chebyshev's theorem can be applied to any distribution regardless of its shape.

Empirical or Normal Rule: Only valid with normal distribution (bell-shaped curve). Approximately 68% lies within 1 standard deviation of the mean; 95% within 2 standard deviations; and 99.7% within 3 standard deviations of the mean.

Standard Score or Z-Score: The value obtained by subtracting the mean and dividing by the standard deviation. When all values are transformed to their standard scores, the new mean (for Z) will be zero and the standard deviation will be one.

Percentile: The percent of the population which lies below that value. The data must be ranked to find percentiles.

Quartile: Either the 25th, 50th, or 75th percentiles. The 50th percentile is also called the median.

Decile: Either the 10th, 20th, 30th, 40th, 50th, 60th, 70th, 80th, or 90th percentiles.

Lower Hinge: The median of the lower half of the numbers (up to and including the median).

Upper Hinge: The median of the upper half of the numbers (including the median). The upper hinge is the 3rd Quartile unless the remainder when dividing the sample size by four is 3.

Box and Whiskers Plot: A graphical representation of the minimum value, lower hinge, median, upper hinge, and maximum.

Inter Quartile Range: The difference between the 3rd and 1st Quartiles.

Outlier: An extremely high or low value when compared to the rest of the values.

Outcome: The result of a single trial of a probability experiment

Sample Space: Set of all possible outcomes of a probability experiment

Event: One or more outcomes of a probability experiment

Equally Likely Events: Events which have the same probability of occurring.

Mutually Exclusive Events / Disjoint Events: Two events which cannot happen at the same time.

Independent Events: Two events are independent if the occurrence of one does not affect the probability of the other occurring.

Dependent Events: Two events are dependent if the first event affects the outcome or occurrence of the second event in a way the probability is changed.

Classical / Theoretical Probability: Uses the sample space to determine the numerical probability that an event will happen.

Empirical Probability: Uses a frequency distribution to determine the numerical probability. An empirical probability is a relative frequency.

Subjective Probability: Uses probability values based on an educated guess or estimate. It employs opinions and inexact information.

Conditional Probability: The probability of an event occurring given that another event has already occurred.

Bayes' Theorem: A formula which allows one to find the probability that an event occurred as the result of a particular previous event.

A **type I error** (level of significance): True null hypothesis is rejected.

A **type II error:** False null hypothesis is accepted.

When fix the level of significance at a predetermined value 0.05 or 0.01. The corresponding confidence levels of test are 95 % and 99% respectively. The 95 % and 99 % confidence intervals for normal distribution are $\mu \pm 1.96\ \sigma$ and $\mu \pm 2.58\ \sigma$ respectively. The coefficient 1.96 and 2.58 are called the 95% and 99% confidences coefficient.

Confidence Intervals: A level of confidence is the probability that the interval estimate will contain the parameter. The level of confidence is 1 - alpha. 1-alpha area lies within the confidence interval.

Estimator: A sample statistic which is used to estimate a population parameter. It must be unbiased, consistent, and relatively efficient.

Interval Estimate: A range of values used to estimate a parameter.

Maximum Error of the Estimate: The maximum difference between the point estimate and the actual parameter. The maximum error of the estimate is 0.5 the width of the confidence interval for means and proportions.

Consistent Estimator: An estimator which gets closer to the value of the parameter as the sample size increases.

Point Estimate: A single value used to estimate a parameter.

Relatively Efficient Estimator: The estimator for a parameter with the smallest variance.

Degrees of Freedom: In statistics, the number of **degrees of freedom** is the number of values in the final calculation of a statistic that are free to vary.

T distribution: A distribution used when the population variance is unknown.

Unbiased Estimator: An estimator whose expected value is the mean of the parameter being estimated.

Probability Distribution: The values a random variable can assume and the corresponding probabilities of each.

Expected Value: The theoretical mean of the variable.

Binomial Experiment: An experiment with a fixed number of independent trials. Each trial can only have two outcomes, or outcomes which can be reduced to two outcomes. The probability of each outcome must remain constant from trial to trial.

Binomial Distribution: The outcomes of a binomial experiment with their corresponding probabilities. Parameter of this distribution are mean = np and variance= npq, it indicate that variance of binomial distribution is always less than the mean.

Multinomial Distribution: A probability distribution resulting from an experiment with a fixed number of independent trials. Each trial has two or more mutually exclusive outcomes. The probability of each outcome must remain constant from trial to trial.

Poisson distribution: A probability distribution used when a density of items is distributed over a period of time. The sample size needs to be large and the probability of success to be small.

Hyper geometric Distribution: A probability distribution of a variable with two outcomes when sampling is done without replacement.

F-distribution: The ratio of two independent chi-square variables divided by their respective degrees of freedom. If the population variances are equal, this simplifies to be the ratio of the sample variances.

Analysis of Variance (ANOVA): A technique used to test a hypothesis concerning the means of three or more populations.

One-Way Analysis of Variance: Analysis of Variance when there is only one independent variable. The null hypothesis will be that all population means are equal, the alternative hypothesis is that at least one mean is different.

Between Group Variation: The variation due to the interaction between the samples, denoted SS(B) for Sum of Squares Between groups. There are k samples involved with one data value for each sample (the sample mean), so there are k-1 degrees of freedom.

Between Group Variance:. This is the between group variation divided by its degrees of freedom.

Within Group Variation: The variation due to differences within individual samples, denoted SS(W) for Sum of Squares Within groups. Each sample is considered independently, no interaction between samples is involved. The degrees of freedom is equal to the sum of the individual degrees of freedom for each sample (df = N – k).

Within Group Variance: The variance due to the differences within individual samples, denoted MS(W) for Mean Square Within groups. This is the within group variation divided by its degrees of freedom.

Scheffe' Test: When the analysis of variance indicates the means are not all equal scheffe test used to find out the differences between means lie. The Scheffe' test is generally used when the sample sizes are different.

Tukey Test: When the analysis of variance indicates the means are not all equal turkry test used to find out the differences between means lie. The Tukey test is generally used when the sample sizes are all the same.

Factors: The two independent variables in a two-way ANOVA.

Treatment Groups: Groups formed by making all possible combinations of the two factors. For example, if the first factor has 3 levels and the second factor has 2 levels, then there will be 3x2=6 different treatment groups.

Interaction Effect: The effect one factor has on the other factor

Main Effect: The effects of the independent variables.

Student unpaired t –test: Difference between means of two independemt samples taken from population with equal variance. Degree of freedom $n_1 + n_2 - 2$.

F-Test

The F-distribution is formed by the ratio of two independent chi-square variables divided by their respective degrees of freedom.

Since F is formed by chi-square, thus many of the chi-square properties carry over to the F distribution.

$$F = \frac{\dfrac{df_1 \cdot s_1^2}{\sigma_1^2} / df_1}{\dfrac{df_2 \cdot s_2^2}{\sigma_2^2} / df_2}$$

- The F-values are all non-negative
- The distribution is non-symmetric
- The mean is approximately 1
- There are two independent degrees of freedom, one for the numerator, and one for the denominator.
- There are many different F distributions, one for each pair of degrees of freedom.

Contingency Table: Data arranged in table form for the chi-square independence test

Expected Frequency: The frequencies obtained by calculation.

Goodness-of-fit Test: A test to see if a sample comes from a population with the given distribution.

Independence Test: A test to see if the row and column variables are independent.

Observed Frequency: The frequencies obtained by observation.

Chi-square distribution: A distribution obtained from the multiplying the ratio of sample variance to population variance by the degrees of freedom when random samples are selected from a normally distributed population

Properties of the Chi-Square

- Chi-square is non-negative and non-symmetric.
- There are many different chi-square distributions, one for each degree of freedom.
- The degrees of freedom when working with a single population variance is n^{-1}.
- the chi-square goodness-of-fit test is always a right tail test.

Multiple Choice Questions

1. The usage of two negatives in the same sentence is called:
   a) **Negation** b) Paralogic
   c) Repression d) Equivocation
2. The first step in summarizing the data is:
   a) Tabulation b) **Classification**
   c) Interpretation d) Observation
3. An interval in which scores are grouped is called:
   a) Point b) Value
   c) **Class interval** d) None of the above
4. The first step in analysis of data is:
   a) Manipulation b) **Categorization**
   c) Interpretation d) Summarization
5. If there is an increase in a series at constant rate, the graph will be:
   a) Convex curve b) Concave curve
   c) Parabola d) **Straight line**
6. A subset or portion of a population is called:
   a) Population b) **Sample**
   c) Representative d) All of the above
7. The population for sampling is defined in terms of:
   a) Time b) Element
   c) Extent d) **All of the above**

8. The key of accurate sampling is:
   a) **Randomness** b) Opinion
   c) Prediction d) All of the above
9. Unrestricted probability sampling method:
   a) Area sampling b) **Simple random sampling**
   c) Quota sampling d) Systematic sampling
10. The units within the between each stratum in a stratified random sampling are respectively:
    a) Homogenous and homogenous
    b) **Homogenous and Heterogeneous**
    c) Heterogeneous and Heterogeneous
    d) Heterogeneous and Heterogeneous
11. The co-efficient value of co-relation lies between:
    a) 0 to 1 b) 0 to -1
    c) -1 to 1 d) 1 - 10
12. Given the N values in a series, the geometric mean is:
    a) The third root of the product of N values.
    b) The square root of the product of N values
    c) The fourth root of the product of N values
    d) **The Nth root of the product of N values**
13. Which of the measures given here are based on every item of the series?
    a) Range b) **Standard deviation**
    c) Quartile deviation d) All of them
14. A researcher selects a probability sample of 100 out of the total population. It is:
    a) A cluster sample **b) A random sample**
    c) A systematic sample d) A stratified sample
15. Generalized conclusion on the basis of a sample is technically known as:
    a) **Statistical inference of external validity of the research**
    b) Parameter inference
    c) Data analysis and interpretation
    d) All of the above

16. A statistical measure based upon the entire population is called:

a) **Parameter** b) Inference

c) Statistic d) None of these

17. If the scatter diagram is drawn and if the points come to a straight line it is an indication of:

a) Skewness **b) Perfect correlation**

c) No correlation d) None of the above

18. If x is 48, Median is 47, Mode value will be:

a) 44 b) 50

c) 45 d) 49

19. In a truly normal frequency distribution:

a) The mean always is the same as the standard deviation

b) The mean is never the same as the mode

c) The mode is never the same as the median

d) The mean always is the same as the median

20. The null hypothesis in an experiment would be:

a) There is a high correlation between the independent and dependent variables

b) **Changing the independent variable has no significant effect on the dependent variable**

c) Changing the dependent variable causes a significant change in the independent variable

d) The standard error of the dependent variable is greater than the mean of the independent variable

21. In general, which of the following statements is FALSE?

a) The sample mean is more sensitive to extreme values than the median.

b) The sample range is more sensitive to extreme values than the standard deviation.

c) The sample standard deviation is a measure of spread around the sample mean.

d) The sample standard deviation is a measure of central tendency around the median.

22. Which of the following statements is NOT true?
    a) In a symmetric distribution, the mean and the median are equal.
    b) The first quartile is equal to the twenty-fifth percentile.
    c) In a symmetric distribution, the median is halfway between the first and the third
    quartiles.
    d) The median is always greater than the mean.
23. What is one of the distinctions between a population parameter and a sample statistic?
    a) A population parameter is only based on conceptual measurements, but a sample statistic is based on a combination of real and conceptual measurements.
    b) A sample statistic changes each time you try to measure it, but a population parameter remains fixed.
    c) A population parameter changes each time you try to measure it, but a sample statistic remains fixed across samples.
    d) The true value of a sample statistic can never be known but the true value of a population parameter can be known.
24. Which one of the following variables is not categorical?
    a) Age of a person.
    b) Gender of a person: male or female.
    c) Choice on a test item: true or false.
    d) Marital status of a person (single, married, divorced, other)
25. Which one of these statistics is unaffected by outliers?
    a) Mean **b) Interquartile range**
    c) Standard deviation d) Range
26. One use of a regression line is:
    a) to determine if any x-values are outliers.
    b) to determine if any y-values are outliers.
    c) to determine if a change in x causes a change in y.
    d) to estimate the change in y for a one-unit change in x.
27. Regression coefficient takes values from:
    a) $-\infty$ to 1 b) 1 to ∞
    c) **$-\infty$ to $+\infty$** d) -1 to +1

28. A statistical parameter, while measure based upon a sample is known as:

a) Parameter
b) Inference
c) Statistic
d) None of these

29. A variable whose value is a numerical outcomes of a random phenomenon is called:

a) a random variable
b) a parameter
c) a random sample
d) a statistic

30. Which of the following provides a measure of central location for the data?

a) Standard deviation
b) Mean
c) Variance
d) Range

31. A numerical value used as a summary mea]sure for a sample, such as sample mean, is known as a:

a) Population parameter
b) Sample parameter
c) Sample statistic
d) Population mean

32. Since the population size is always larger than the sample size, then the sample statistic:

a) Can never be larger than the population parameter
b) Can never be equal to the population parameter
c) Can be smaller, larger, or equal to the population parameter
d) Can never be smaller than the population parameter

33. 'μ' is an example of a:

a) Population parameter
b) Sample statistic
c) Population variance
d) Mode

34. The hourly wages of a sample of 130 system analysts are given as mean = 60 range = 20 mode = 73 variance = 324 median = 74. The coefficient of variation equals:

a) 0.30%
b) 30%
c) 5.4%
d) 54%

35. The variance of a sample of 169 observations equals 576. The standard deviation of the sample equals:

a) 13
b) 24
c) 576
d) 461

36. The median of a sample will always equal the:

a) Mode
b) Mean
c) 50th percentile
d) All of the above

37. The median is a measure of:

a) Relative dispersion
b) Absolute dispersion
c) Central location
d) Relative location

38. The 75th percentile is referred to as the:

a) First quartile
b) Second quartile
c) Third quartile
d) Fourth quartile

39. The difference between the largest and the smallest data values is the:

a) Variance
b) Interquartile range
c) Range
d) Coefficient of variation

40. The first quartile:

a) Contains at least one third of the data elements
b) Is the same as the 25th percentile
c) Is the same as the 50th percentile
d) Is the same as the 75th percentile

41. Which of the following is not a measure of central location?

a) Mean
b) Median
c) Variance
d) Mode

42. If a data set has an even number of observations, the median:

a) Cannot be determined
b) Is the average value of the two middle items
c) Must be equal to the mean
d) Is the average value of the two middle items when all items are arranged in ascending order

43. Which of the following is a measure of dispersion?

a) Percentiles
b) Quartiles
c) Interquartile range
d) All of the above

44. The most frequently occurring value of a data set is called the:

a) Range
b) Mode
c) Mean
d) Median

45. The interquartile range is:

a) The 50th percentile
b) Another name for the variance

c) The difference between the largest and smallest values

d) The difference between the third quartile and the first quartile

46. When the data are skewed to the right, the measure of skewness will be:

a) Negative
b) Zero
c) Positive
d) One

47. When data are positively skewed, the mean will usually be:

a) Greater than the median
b) Smaller than the median
c) Equal to the median
d) Positive

48. Which of the following is **not** a measure of dispersion?

a) The range
b) The 50th percentile
c) The standard deviation
d) The interquartile range

49. The interquartile range is used as a measure of variability to overcome what difficulty of the range?

a) The sum of the range variances is zero
b) The range is difficult to compute
c) The range is influenced too much by extreme values
d) The range is negative

50. If the variance of a data set is correctly computed with the formula using n - 1 in the denominator, which of the following is true?

a) The data set is a sample
b) The data set is a population
c) The data set could be either a sample or a population
d) The data set is from a census

51. In computing descriptive statistics from grouped data:

a) Data values are treated as if they occur at the midpoint of a class
b) The grouped data result is more accurate than the ungrouped result
c) The grouped data computations are used only when a population is being analyzed
d) None of these alternatives is correct.

52. The measure of dispersion that is influenced most by extreme values is:

a) The variance
b) The standard deviation
c) The range
d) The interquartile range

53. When should measures of location and dispersion be computed from grouped data rather than from individual data values?

a) As much as possible since computations are easier

b) Only when individual data values are unavailable

c) Whenever computer packages for descriptive statistics are unavailable

d) Only when the data are from a population

54. The descriptive measure of dispersion that is based on the concept of a deviation about the mean is:

a) The range

b) The interquartile range

c) The absolute value of the range

d) The standard deviation

55. The measure of location which is the most likely to be influenced by extreme values in the data set is the:

a) Range

b) Median

c) Mode

d) Mean

56. The most important statistical descriptive measure of the location of a data set is the:

a) Mean

b) Median

c) Mode

d) Variance

57. The numerical value of the standard deviation can never be:

a) Larger than the variance

b) Zero

c) Negative

d) Smaller than the variance

58. The sample variance:

a) Is always smaller than the true value of the population variance

b) Is always larger than the true value of the population variance

c) Could be smaller, equal to, or larger than the true value of the population variance

d) Can never be zero

59. The variance can never be:

a) Zero

b) Larger than the standard deviation

c) Negative

d) Smaller than the standard deviation

60. If two groups of numbers have the same mean, then:
   a) Their standard deviations must also be equal
   b) Their medians must also be equal
   c) Their modes must also be equal
   d) None of these alternatives is correct

61. The sum of deviations of the individual data elements from their mean is:
   a) Always greater than zero
   b) Always less than zero
   c) Sometimes greater than and sometimes less than zero, depending on the data elements
   d) Always equal to zero

62. Which of the following symbols represents the standard deviation of the population?
   a) s^2 **b) s**
   c) μ d) x

63. Which of the following symbols represents the mean of the population?
   a) s^2 b) s
   c) μ d) x

64. Which of the following symbols represents the variance of the population?
   a) s^2 b) s
   c) μ d) x

65. Which of the following symbols represents the size of the population?
   a) s^2 b) s
   c) μ **d) N**

66. Which of the following symbols represents the mean of the sample?
   a) s^2 b) s
   c) μ **d) x**

67. Which of the following symbols represents the size of the sample:
   a) s^2 b) s
   c) N **d) n**

68. The symbol 's' is used to represent:
   a) Variance of the population
   b) The standard deviation of the sample
   c) Standard deviation of the population
   d) The variance of the sample
69. The symbol s2 is used to represent:
   a) The variance of the population
   b) The standard deviation of the sample
   c) The standard deviation of the population
   d) The variance of the sample
70. A numerical measure of linear association between two variables is the:
   a) Variance
   b) Covariance
   c) Standard deviation
   d) Coefficient of variation
71. Positive values of covariance indicate:
   a) A positive variance of the x values
   b) A positive variance of the y values
   c) The standard deviation is positive
   d) Positive relation between the independent and dependent variables
72. A numerical measure of linear association between two variables is the:
   a) Variance
   b) Coefficient of variation
   c) Correlation coefficient
   d) Standard deviation
73. The coefficient of correlation ranges between:
   a) 0 and 1
   b) -1 and +1
   c) Minus infinity and plus infinity
   d) 1 and 100
74. The coefficient of correlation:
   a) Is the same as the coefficient of determination
   b) Can be larger than 1
   c) Cannot be larger than 1
   d) Cannot be negative
75. When the smallest and largest percentage of items is removed from a data set and the mean is computed, the mean of the remaining data is:
   a) The median
   b) The mode
   c) The trimmed mean
   d) Any of the above

76. In a five number summary, which of the following is **not** used for data summarization?
    a) The smallest value
    b) The largest value
    c) The mean
    d) The 25th percentile

77. Which of the following is **not** a measure of dispersion?
    a) Mode
    b) Standard deviation
    c) Range
    d) Interqurtile range

78. Since the mode is the most frequently occurring data value, it:
    a) Can never be larger than the mean
    b) Is always larger than the median
    c) Is always larger than the mean
    d) None of these alternatives is correct.

79. Degree of freedom of x^2 test for 4 x 3 contingency table is:
    a) 12
    b) **6**
    c) 1
    d) 7

80. Statistical conclusions are always:
    a) Absolutely true
    b) Absolutely wrong
    c) Either of a or b
    d) **True on average**

81. Harmonic mean is better than other means if data are for:
    a) **Speeds or rates**
    b) Heights and lengths
    c) Binary values 0 and 1
    d) Ratio and proportion

82. Harmonic mean gives weight age to:
    a) **Small values**
    b) Positive values
    c) Large values
    d) Negative values

83. To study the average height of plants, we use:
    a) Mode
    b) Harmonic mean
    c) **Arithmetic mean**
    d) Geometric mean

84. Which among the following is not a measure of central tendency:
    a) Mean
    b) Median
    c) Mode
    d) Variance

85. Mean deviation is minimum when taken about:
    a) Mean
    b) **Median**
    c) Mode
    d) Variance

86. The relation between arithmetic mean, geometric mean and harmonic mean is:

a) AM=GM=HM b) **AM>GM>HM**

c) AM<GM<HM d) AM<HM<GM

87. Range is not a reliable measure of:

a) Mean b) **Dispersion**

c) Median d) Mode

88. Formula for calculating coefficient of variation is:

a) **SD/Mean x 100** b) Mean/SD x 100

c) Mean x SD/100 d) 100/Mean x SD

89. The degree of linear relationship between variables are expressed through:

a) **Coefficient of correlation** b) Standard deviation

c) Arithmetic mean d) None of the above

90. Regression coefficient is independent of:

a) **Origin** b) Scale

c) Both origin and scale d) Neither origin nor scale

91. Mean and variance of a standard normal distribution are respectively:

a) **0 and 1** b) Geometric mean

c) Both a and b d) None of the above

92. The degrees of freedom for statistics for paired t test of n pairs of observations is:

a) 2 (n -1) b) **n – 1**

c) 2n – 1 d) None of these

93. Hypothesis that the population variance has a specified value can be tested by:

a) F test b) Z test

c) **χ^2 test** d) t test

94. For testing the independence of two attributes, we use:

a) F test b) Z test

c) **χ^2 test** d) t test

95. The value of x^2 always lies between:

a) 0 to 1 b) **0 to infinity**

c) -1 to +1 d) - infinity to + infinity

96. For testing the significance of correlation coefficient, we use:

a) F test
b) ANOVA
c) x^2 test
d) **t test**

97. The parent population in analysis of variance is assumed to follow:

a) **Normal distribution**
b) Binomial distribution
c) Poisson distribution
d) None of the above

98. Who is considered as father of statics:

a) **Ronald fisher**
b) Emile Dukeman
c) Francis Galton
d) Carl pearson

99. The non-parametric substitute for the parametric 't' test:

a) X^2 test
b) Kvuskal Wallis test H test
c) Friedman test
d) **Mann-Whitney U test**

100. An example for one way non-parametric analysis of variance among the following is:

a) F test
b) **Kvuskal Wallis test**
c) Friedman test
d) Z test

101. The level of measurement associated with Man-Whitney U test:

a) Nominal
b) **Ordinal**
c) Interval
d) Ratio

102. Pick the odd man out:

a) F test
b) Kvuskal Wallis test
c) Friedman test
d) **Z test**

103. Greater the distance of Lorenz curve from the line of equal distribution:

a) **More is inequality**
b) Less is inequality
c) Nothing can be inferred
d) None of the above

104. A normal curve is all but not:

a) Symmetrical
b) Continuous
c) Unimodel
d) **Asymbiotic**

105. The value of probability ranges from:

a) -1 to +1
b) 0 to infinity
c) **0 to 1**
d) None of the above

106. Probability of a leap year will year have 53 Sunday is:
   a) 1/7 b) **2/7**
   c) 2/53 d) 52/53
107. The probability of an impossible event is:
   a) **0** b) 1
   c) Infinity d) None of the above
108. The basis of sampling theory is:
   a) Probability b) **Randomization**
   c) Local control d) Relative Frequency
109. List of sampling units is:
   a) Sample population b) **Sampling frame**
   c) Sample element d) Sampling unit
110. Difference between estimated value and actual is measured as:
   a) Non sampling error b) Bias
   c) **Sampling error** d) None of the above
111. Level of significance is the probability of:
   a) **Type I error** b) Type II error
   c) Not committing error d) Any one of the above
112. Rejecting a hypothesis when it is true:
   a) **Type I error** b) Type II error
   c) Not committing error d) Any one of the above
113. Area of critical region depends upon:
   a) **Size of type I error** b) Size of type II error
   c) Both of the above d) None of the above
114. Accepting a hypothesis when it is false:
   a) Type I error b) **Type II error**
   c) Both a and b d) None of the above

11

Overview of Adult Education (Theory)

Philosophy Of Adult Education

Philosophy	Liberal	Behaviorist	Progressive	Humanistic	Radical
Purpose	To make a person liberate in broadest sense: intellectually, morally, spiritually	To control the behavior of a person within an organization	To transmit culture, promote social change, to reform society	To enhance personal and development facilitate actualization	To bring fundamental social, political and economic changes in society,
Learner	"Renaissance" person cultural and always a keen learner.	Active role player in packaged and programmed learning	People who have unlimited potential to be developed through problem solving	Assumes responsibility for learning and self development.	Equality with teacher in learning process personal autonomy.
Teacher	'Expert' who transmits knowledge. Authoritative To down, vertical com-munication.	Manager and controller. Predicts and direct learning outcome.	Organizer guides learning through experience.	Facilitator, partner promotes but did not guide.	Provocateur Equality between teacher and learner
Source of Authority	Western canons of Liberty and individualism	Environment Organization Society	Situations Cultural and Social experience.	The self learner	Socio-economic Socio- political
Key concepts	Learning for its own sake	Stimulus- Response Behaviour modification	Problem solving experience based	Experiential learning self directedness	Conscious-ness raising praxis
Methods	Didactic, Monologue lecture	Programmed instruction	Problem solving	Experiential group tasks	Maximum interaction
Time Frame	Oldest Philosophy, Ancient Greece	J.B. Watson Found in 1920	Pragmatism In 1870 USA Serious movement Since Dewey 1900	1950-1960 Maslow, Rogers	Anarchist

Different Theories of Adult Learning

1. **Learning by association** (E.L. Thorndike) - Called as Operant or Instrumental Conditioning. He stressed the importance of reward in forming association. Early Behaviorist has conceived teaching as manipulation of environment to produce desirable change.
2. **Learning by conditioning** (Skinner) propagated this theory related to stimulus response behavior.

- There are two types of conditioning

1. Respondent conditioning
2. Operant conditioning

- Pavlov's experiment, Natural Stimulus got the properties of the Un-Conditioned Stimulus. This is called Respondent Conditioning (Classifiable).
- Skinner Operant Conditioning a response occurs spontaneously in the absence of any specifically correlated stimulus.

3. **Learning by doing** (E.L. Thorndike) Modern psychologist considers the term "Trail & Error". This type of learning is well exemplified by the classical experiment performed by American psychologist E.L. Thorndike. Learning is both a process and product.

- Law of effect: - Response on the stimulus determine it will recur or not.
- Law of exercise: - Strengthening the connection through practice.
- Law of readiness: - Learner's willingness to make the S-R connections.

4. **Learning by insight:** Cognitive theorists have investigated and shown that people learn by perceiving, comprehending and conceptualizing the problem. Based on problem solving approach. Gestalt Psychologist, Kohler conducted experiments on Chimpanzees, restructuring of cognitive field.

Principles of Teaching Adult

1. Adults need to be involved in planning & evaluation
2. Experience provides the basis of learning.
3. Interested in subject of immediate relevance.
4. Problem centered rather than content oriented.

Adult Learner

- Has an independent self-concept?
- Has accumulated a reservoir of life experience
- Learning related to changing social roles
- Problem centered learning
- Motivated to learn internally.

Characteristics

- Problem centred
- Result oriented
- Self directed
- Sceptical about new information
- Accept responsibility for their own learning

Neo literate and Semi-Literate

Neo-Literate: Adult and adolescent who did not make use of available educational opportunities on time.

And

At a later stage acquired the skills of Literacy through formal or non formal approach.

Semi-Literate: Adult and adolescent who have lower level of literacy skill. School drop-outs or who have studied up to 5th standard.

Different Committee on Adult Education

1. **Ashok Mehta Committee:** In December 1977, the Janata Government appointed a committee on Panchayati Raj institutions under the chairmanship of Ashoka Mehta. The committee submitted its report in August 1978 and made 132 recommendations to revive and strengthen the declining Panchayti Raj system in the country. The main recommendations of the committee were:

- The 3-tier system of Panchayati Raj should be replaced by the 2-tier system: Zilla Parishad at the district level, and below it, the Mandal Panchayat consisting of a group of villages convering a population of 15000 to 20000.
- A district should be the first point for decentralistion under popular supervision below the state level.

- Zila Parishad should be the executive body and made responsible for planning at the district level.

2. **Balwant Rai Mehta Committee:** Balwant Rai G. Mehta Committee was appointed in 1957 to make recommendations for the revitalization of the Panchayati Raj system and define its role in the development process. The report of the Committee recommended that public participation in community works should be organized through statutory representative bodies. The specific recommendations of the committee were establishment of a 3-tier Panchayati Raj system – Gram Panchayat at the village level, Panchayat Samiti at the block level, and District Praishad at the district level. These tiers should be organically linked through a device of indirect elections.

3. **Narasimhan Committee:** In 1971, the Andhra Pradesh Government appointed a High Power Committee under the Chairmanship of C. Narasimham, a retired I.A.S. Officer to examine the working of the Panchayati Raj and made proposals for reorganizing it. The committee submitted its report in 1972. Some of the important recommendations made by the committee are: the member of Legislative Assembly and members of Parliament should be completely kept out of Panchayati Raj system, the Panchayati Raj system be modified on the Maharashtra pattern, the District Collector should be kept out of the Panchayati Raj and instead another I.A.S. Officer should be appointed at the chief executive officer of the Zilla Parishad. The Sarpanchof the Village Panchayat will be elected directly by the people, the members of Zilla Parishads should also be directly elected by the people.

4. **Vengal Rao Committee:** Under the headship of J. Vengala Rao consisting of 13 members of Legislative Assembly was set up in 1968 to consider all aspects relating to the panchayati Raj and to suggest improvements and amendments to the concerned Acts. Some recommendations of the Committee were:

- Reforms to rationalize the mode and pattern of elections to the Panchayati Raj Institutions,
- Reduction of the importance of the village Sarpanch in the Panchayati Raj set up
- Giving up the provision for Co-operation of Village panchayats, Panchayat Samitis or zilla parishads,
- Giving representation to Scheduled Castes, Scheduled Tribes and women in the Standing Committee by means of co-option,

- Making rule by standing committees, mandatory Sarpanch the samithi President or the Zilla Parishad Chairman will have no independent power but would act strictly accordingly to Standing Committee decision.

5. **Yashpal Committee:** In the year 2009, the Ministry of Human Resource Development (MHRD) had set up a Committee on Higher Education known as Yashpal Committee. It suggested the scrapping of all higher education regulatory/monitoring bodies and creation of a super regulator: a seven-member Commission for Higher Education and Research (CHER).
6. **M.S Mehta Committee:** The Education Commission (1964-66) recommended that a vigorous effort be made to establish specialized institutions to provide vocational education in agriculture and allied fields at the pre and post matriculate levels to cater the training needs of a large number of boys and girls coming from rural areas. The Indian Council of Agricultural Research (ICAR), therefore, constituted a committee in 1973 headed by Dr. Mohan singh Mehta of Seva mandir, Udaipur (Rajasthan), for working out a detailed plan for implementing this scheme, The Committee submitted its report in 1974. The first Krishi Vigyan Kendra (KVK), on a pilot basis, was established in 1974 at Puducherry (Pondicherry) under the administrative control of the Tamil Nadu Agricultural University, Coimbatore.

Multiple Choice Questions

1. Andragogy and pedagogy basically both refers to:
   a) Understanding suitable teaching strategies
   b) Understanding the learning process
   c) Understanding learner's need
   d) Understanding teaching process
2. Adult Education programme includes:
   a) Functional Literacy
   b) Adjustment and preparation of world work.
   c) Agent working as a bridge between younger generation and older generation.
   d) All the above
3. National literacy mission in 1988 was launched with
   a) Four Technology Missions
   b) Five Technology Missions
   c) Four Educational Missions
   d) Three Technology Missions

4. In which year, National open school was established:
   a) 1970
   b) 1969
   c) 1971
   d) 1972
5. The main task of an adult education programmer is :
   a) To develop interest in education
   b) To identify reasons for resistance to programmed.
   c) To impart functional literacy for Quality of life improvement
   d) None of the above
6. The main office of National open school in India is:
   a) Mumbai
   b) Hyderabad
   c) Chennai
   d) Delhi
7. 'Nehru Literacy Award' was constituted by agency:
   a) IAEA
   b) IUACE
   c) Ministry of HRD
   d) Directorate of Adult Education
8. The concept of Social Education was initiated during:
   a) First Five Year Plan
   b) Second Five Year Plan
   c) Third Five Year Plan
   d) Fourth Five Year Plan
9. The National Adult Education Programme was started in the year:
   a) 1975
   b) 1976
   c) 1977
   d) 1978
10. Denmark, Adult Education movement was known as:
    a) Folk High Schools
    b) Grammer Schools
    c) Public Schools
    d) Residential Schools
12. Area Development Approach depends on:
    a) Major Planning
    b) Minor Planning
    c) Micro Planning
    d) Macro Planning
13. German Adult – Education is known as:
    a) GAEA
    b) GEA
    c) DVV
    d) GUAE
14. 'Animator' in Adult Educational Programme are:
    a) Facilitators
    b) Communicators
    c) Teachers
    d) Researchers

15. National Adult Education Programme defined adult education as :
    a) Literacy, Functionality and Conscientization
    b) Functional Literacy, Productivity and Awareness
    c) Mass Education, Fundamental Education, Basic Education
    d) Liberal Education, Literacy and Education For Women
16. Insight Learning' is associated with:
    a) Kohler b) Maslow
    c) Thorndike d) Skinner
17. Population Education mean:
    a) Education about the different type4s of people in different countries
    b) Education about the impact of a large pop0ulation on development
    c) Education about family planning and contraceptives
    d) Education about the reasons for the population growth in India
18. Participatory Research aim at:
    a) Developing the discipline of adult education
    b) Solving problems of people
    c) Posing question of ethics of research
    d) Mobilizing and empowering community members to develop and conduct research and use research outputs
19. International Literacy day is celebrated on:
    a) 18th July **b) 8th September**
    c) Thorndike d) 19th November
20. Paradigm Shift in 'POPED' includes:
    a) Reproductive Health b) Small Family Norms
    c) Favorable Sea Ratio d) Crude Birth Rate
21. Population density is measurement of:
    a) Types of Houses where people live
    b) Number of people in per square k.m.
    c) Distribution of population in the area
    d) None of the above

22. The three key components of development as indicated by UNDP reports are:

    a) Adult Literacy, Education, Income

    b) Longevity, Knowledge and Income

    c) Health Care, Home, Intelligence

    d) Hygiene, Nutrition, Income

23. World Declaration on "Education for all" was made in:

    a) 1990 b) 1996

    c) 1980 d) 1992

24. International Population Day is celebrated on:

    a) 18 July **b) 11 July**

    c) 20 July d) 10 July

25. Life Long Education is regarded as:

    a) Education that does not end with schooling but it is a lifelong process

    b) Shifting from formal to non-formal

    c) Life Skill Education for better health

    d) Education for parent hood

26. The first Total Literate District was:

    a) Chittoor **b) Ernakulam**

    c) Ajmer d) Bankura

27. Education Technology is best defined as:

    a) Use of gadgets

    b) Developing Learning Material

    c) Systematic Applications of Learning theories in Teaching

    d) Systematic Applications of Technology in Education

28. The term "Child labour" refers to:

    a) Any work by children b) Only studying

    c) Mental work **d) Destructive work**

29. The Indian Adult Education Association instituted the 'Nehru Literacy Award' in:

    a) 1962 **b) 1965**

    c) 1968 d) 1969

30. National Literacy Mission includes:
   a) Each One Teach One
   b) Farmer's Functional Literacy
   c) New Area Development Approach
   d) Total Literacy campaign
31. Total Literacy Campaign was related only to:
   a) Teaching Adults
   b) 3 R's
   c) Basic Literacy
   d) Skill Development
32. What is the other name of Polyvalent Education Centre?
   a) Shramik Vidyapeeth
   b) Gram Shikshan Mohim
   c) Jan Shikshan Sansthan
   d) Community Education Centre
33. The World Conference on Education for all is held in:
   a) Thailand
   b) Singapore
   c) India
   d) China
34. Population studies are specifically concerned with:
   a) Prevention of Infant Mortality Rate
   b) Democracy
   c) Demography
   d) Quality of life
35. Who has Advocated "Rural Universities will support the Basic Education"?
   a) Sarojini Naidu
   b) Sardar Patel
   c) Jawaharlal Nehru
   d) Mahatma Gandhi
36. Frank LauBach contribution to adult education is :
   a) Formal Education Approach
   b) Non-Formal Education Approach
   c) Each one Teach one Method
   d) Incidental Education Approach
37. "Human Rights Day" is celebrated on:
   a) 10th November
   b) 10th December
   c) 10th July
   d) 10th September
38. Arrange the following in the order in which they appeared. Use the code given below.
   i) Literate
   ii) Illiterate
   iii) Semiliterate
   iv) Neo-literate
   a) ii, i, iii, iv
   b) iv, iii, ii, i
   c) i, ii, iii, iv
   d) ii, iv, i, iii

39. Integrated approach to attain the higher level of literacy means
    a) To use audio – visual system
    b) To remove illiteracy
    c) To cover backward areas
    d) To cover drop out children
40. "SRC" stands for
    a) Space Research Centre
    b) State Re-organization Commission
    c) State Resource Centre
    d) State Research Centre
41. "Folk Schools" are located in
    a) Africa **b) Denmark**
    c) England d) West Indies
42. Billions of people in developing countries are
    a) Feeling miserable
    b) Earning good income
    c) Hoping for the future
    d) Feel positive about development
43. University – community interaction is a
    a) Two way process
    b) One way process
    c) Technology driven process
    d) None of the above
44. Lifelong education is an expansion of the scope of
    a) Adult education b) Population education
    c) legal education **d) Continuing education**
45. Environment pollution is linked to
    a) Global warming b) Ozone depletion
    c) Green house gases **d) All of the above**
46. How many fundamental steps Gandhi's 'Basic Education System' contained?
    a) Three b) Four
    c) Five d) Six

47. Which form of research is done to solve immediate problem?
    a) Action Research b) Pure Research
    c) Survey Research d) Evaluating Research
48. Kalike Galige Programme developed in the State of
    a) Andhra Pradesh b) Telangana
    c) Karnataka d) Maharashtra
49. Which of the following political leader was known as an adult educator?
    a) F. Casfro b) M. Gorvachev
    c) Z.A. Bhutto **d) J. Neyrere**

Some Important Technologies

TBT Agreement- The WTO Agreement on Technical Barriers to Trade (the "TBT Agreement") establishes rules and procedures regarding:

- Development, adoption, and application of voluntary product standards,
- Mandatory technical regulations,
- Procedures (such as testing or certification)
- The Technical Barriers to Trade (TBT) Agreement aims to ensure that technical regulations, standards, and conformity assessment procedures are non-discriminatory and do not create unnecessary obstacles to trade.
- At the same time, it recognises WTO members' right to implement measures to achieve legitimate policy objectives, such as the protection of human health and safety, or protection of the environment.
- The TBT Agreement strongly encourages members to base their measures on international standards as a means to facilitate trade.
- Through its transparency provisions, it also aims to create a predictable trading environment.

Animal Welfare Board of India

- The Animal Welfare Board of India, the first of its kind to be established by any Government in the world, was set up in 1962, in accordance with Section 4 of the Prevention of Cruelty to Animals Acts 1960.
- Shrimati Rukmini Devi Arundale pioneered the setting up of the Board, with its Headquarters at Chennai. She guided the activities of the Board for nearly twenty years till her demise in 1986.
- People for the Ethical Treatment of Animals (PETA)
- Society for Prevention of Cruelty to Animals (SPCA)
- Animal Welfare Organization (AWO)

Animal Birth Control programme

- Many small towns are taking away dogs and killing them because they do not have sterilization. This needs to be stopped immediately by getting the ABC (Animal Birth Control) programme started in your own area.
- This ABC (Animal Birth Control) programme has been developed by WHO as the only practical solution to control the street dog population and eradicate rabies. In cities where it has been wholeheartedly implemented dog numbers and bites have reduced and these are now zero rabies zones.
- Based on this success, GOI has formulated the Dog Rules 2001 which directs that municipalities work with animal welfare organisations to implement the ABC programme.

Hazard Analysis and Critical Control Points (HACCP)

- A systematic preventive approach
- To food safety from biological, chemical, and physical hazards in production processes
- Hazard Analysis and Critical Control Point (HACCP) is an internationally recognized system for reducing the risk of safety hazards in food.
- A HACCP System requires that potential hazards are identified and controlled at specific points in the process. This includes biological, chemical or physical hazards.
- Any company involved in the manufacturing, processing or handling of food products can use HACCP to minimize or eliminate food safety hazards in their product.

Seven principle of HACCP

1. Conduct a Hazard Analysis
2. Identify the Critical Control Points (CCP)
3. Establish a maximum or minimum limit for CCP (temperature, time, pH, salt level, chlorine level or other processing characteristic that will control the hazard).
4. Establish Critical Limits
5. Establish Monitoring Procedures
6. Establish Corrective Actions
7. Establish Record Keeping Procedures

National Programme for Bovine Breeding and Dairy Development: GOI funded project for 12th Five year Plan Period (2013-17). Two components i.e., National Plan for Bovine Breeding (NPBB) & National Plan for Dairy Development (NPDD) with total Outlay (till EOP): Rs. 1800 crore.

Livestock biodiversity: The 'livestock' term refers to animals domesticated for producing commodities for man such as food, fiber and draught. Livestock biodiversity is integral to our culture, history, environment, and economy. Thousands of livestock breeds have evolved over time to suit particular environments and farming systems.

Analytic Hierarchy Process: The Analytic Hierarchy Process (AHP) is a method for organizing and analyzing complex decisions, using math and psychology. It was developed by Thomas L. Saaty in the 1970s and has been refined since then. It contains three parts: the ultimate goal or problem you're trying to solve, all of the possible solutions, called alternatives, and the criteria you will judge the alternatives on. AHP provides a rational framework for a needed decision by quantifying its criteria and alternative options, and for relating those elements to the overall goal.

Structural equation modeling: is a multivariate statistical analysis technique that is used to analyze structural relationships. This technique is the combination of factor analysis and multiple regression analysis, and it is used to analyze the structural relationship between measured variables and latent constructs. This method is preferred by the researcher because it estimates the multiple and interrelated dependence in a single analysis. In this analysis, two types of variables are used endogenous variables and exogenous variables. Endogenous variables are equivalent to dependent variables and are equal to the independent variable.

References

Adams J.S. (1963) Towards An Understanding of Inequality. Journal of Abnormal and Normal Social Psychology. 67, 422-436.

Alderfer, Clayton P. (1969). An Empirical Test of a New Theory of Human Needs; Organizational Behaviour and Human Performance, 4(2), 142–175.

Annual Report (2004-05). Department of Women and Child Development, GOI.

Berlo D.K. (1960). The process of Communication. Holt, Rinehart Winston Inc. New York.

Chitamber J.B. (1997). Introductory Rural Sociology: A synopsis of Concepts and Principles. New age International Limited Publishers, New Delhi.

Colvin J.W. (2007). Peer tutoring and social dynamics in higher education. Mentoring & Tutoring: Partnership in Learning, 15(2), 165–181.

Daivadeenam P. and Somani L.L. (2013). Research Methodology in Extension Education. Agrotech Publishing Academy. 2nd edition. ISBN 108183213014

Department of Women and Child Development (2004). Annual Report, Part IV, Ministry of Human Resource development, Govt. of india, New Delhi.

Directorate of Extension (1961). Extension Education and Community Development. Govt. of India, New Delhi.

Draft of Five year plan (2010). Planning Commission of India. New Delhi.

Drucker, Peter F. (1985). Innovation and Entrepreneurship: attributes the coining and defining of "entrepreneur". A Treatise on Political Economy.

Edward L. Thorndike (1999). Education Psychology: briefer course, New York: Routledge, ISBN 0-415-21011-9

Edward W. Weidner (1963). Development administration: an annotated bibliography East-West Center Press.

Fayol H. (1949). General and Industrial Management. (C. Storrs, Trans.). London: Sir Isaac Pitman & Sons, LTD. (Original work published 1918)

Friedmann J. (1992). Empowerment: the Politics of Alternative Development, Oxford, Basil Blackwell.

Gulick L. and Urwick L. (1937). Papers on the Science of Administration (pp. 3–35). New York: Institute of Public Administration.

Henry L Gantt. (1919). Organizing for work. www.ganttcharts.com.

Herzberg F. (1964). "The Motivation-Hygiene Concept and Problems of Manpower". Personnel Administrator (27), 3–7.

Horton P.B. and Hunt C.L. (1976). Sociology, MacGraw Hill, Tokyo. Inkeles Alex.

http://planningcommission.nic.in/plans/planrel/fiveyr/welcome.html

Kelsey. L.D. and Hearne C.C. (1967). Cooperative Extension Work. Cornell University Press, Ithaca. New York.

Kerlinger F.N. (1966). Foundation of Behavioural Research. 3rd edition Prism Book Pvt. Ltd. Banagalore.

Kolb D.A. (1984). Experiential learning: experience as the source of learning and development Englewood Cliffs, NJ: Prentice Hall. URL: http://www.learningfromexperience.com/images/uploads/process-of-experiential-learning.pdf

Koontz H., and Weihrich H. (1988). Industrial management McGraw-Hills. 9thedition. ISBN-0071005420, 9780071005425

Kotler Philip. (1967). Marketing Management: Analysis, Planning and Control. Englewood Cliffs, N.J.: Prentice-Hall.

Leagans J.P. (1967). A Concept of the Extension Education Process in Cooperative Extension Work, (eds.) Cornell University Press, Ithaca, New York.

Leagans J.P. (1961). Characteristics of Teaching and Learning in Extension Education, Extension Teaching Methods, The Communication Process and Programme Planning to Meet People's Needs, in Extension Education in Community Development, Directorate of Extension, Govt. of India, New Delhi.

Leagans J.P. (1971). Extension education and modernization. In: Leagans, J.P. & Loomis, P.L. (eds). Behavioural Change in Agriculture, Concepts and Strategies for Influencing Transmission. pp. 101–47. Cornell Univ. Press, USA.

Legans J.P. (1961). The Communication Process, Extension Education in Community Development, Directorate of Extension, Ministry of Food and Agriculture, Govt. of India, New Delhi

Likert R. (1961). New patterns of Management. McGraw-Hill Book Company, New York.

Louis E. Boone and David L. Kurtz. (2006). Contemporary Marketing Thomson South-Western, 12 th edition 10: 0324359292

Lundberg G.A. (1968). Sociology 4th edition., McGraw Hill.

MacIver R.M., and Page C.H. (1977). Society: An Introductory Analysis. The MacMillan Co. of India Ltd., Mumbai.

McClelland D.C. (1961). The Achieving Society. Princeton, NJ: Van Nostrand. ISBN 978-0029205105.

Maslow A.H. (1943). A Theory of Human Motivation. Psychological Review, 50(4), 370-396.

Mathialagan P. (2007). Text Book of Animal Husbandry & Livestock Extension. International Book Distributing Co. 3rd edition. ISBN 8181892100

McGregor D. (1960). The human side of enterprise.1st edition. McGraw-Hill, New York.

Ministry of Rural Development (2001). Swarnjayanti Gram Swarojgar Yojna. Kurukshetra, November.

O.P. Dhama. (1989). Rural welfare, Ram Prasad & Sons, Agra.

Ralph Linton. (1954). Cultural Background of Personality.

Rappaport J. (1987). Terms of empowerment/exemplars of prevention: Toward a theory for community psychology American Journal of Community Psychology, 15(2), 121-148.

Ray G.L. (2011). Extension Communication and Management. Kalyani Publishers. 8th edition. ISBN 9788127268404

Robinson F.P. (1961). Effective study Publisher: Harper; Revised edition (1961). ASIN: B0006AWXRC

Rogers E.M. and Shoemaker F.F. (1971). Communication of Innovations. The Free Press, New York.

Rousseau J.J. (1953). The Stanford Encyclopedia of Philosophy (Winter 2012 Edition), http://plato.stanford.edu/archives/win2012/entries/rousseau/>.

Sandhu A.S. (1994). Extension Programme Planning. Oxford and IBH publishing Co. Pvt. Ltd., New Delhi. ISBN 8120409116.

Saravanam R., Kathiresan C. and Indradevi T. (2011). Information and Communication Technology for Agriculture and Rural Development. New India Publishing Agency, New Delhi. 1st edition ISBN 9789380235882

Schramm W. (1961). The Process and Effects of Mass Communication. University of Illinois Press, Illinois.

Shanon C. and Wever W. (1949). The Mathematical Theory of Communication. University of Illinois Press, Illinois.

Skinner B.F. (1953). Science and Human Behavior. ISBN 0-02-929040-6

Stevens S.S. (1946). Theory of Scales of Measurement Science, New Series, Vol. 103, No. 2684, pp. 677-680 American Association for the Advancement of Science Stable URL: http://www.jstor.org/stable/1671815

Sunil.V.G. (2008). Agriculture Extension, 3rdedition. New Vishal Publication, New Delhi. ISBN-8183990762

Sutherland Lee R. (1961). Introductory Sociology. J.B. Lippincott Publisher.

UNDP (1995). Human Development Report. Published for the United Nations Development Programme (UNDP) New York Oxford University Press 1995ISBN 0-19-510023-9

UNESCO'S (2002-07). Gender Mainstreaming Implementation Framework (GMIF) for http://agritech.tnau.ac.in/atma/atma_intro.html

Vroom Victor H. (1964). Work and motivation. John Wiley & Sons, Inc.

Watson J.B. (1919). Psychology from the standpoint of a behaviorist. Philadelphia: Lippencott.

Wells A.F. (1936). The Sociological Review. Volume a28, Issue 3, pages 274–294, Article first published online: 18 AUG 2011, DOI: 10.1111/j.1467-954X.1936.tb01333.x

Westley and MacLean (1957). A Conceptual Model I for Communications Research Downloaded from jmq.sagepub.com

Wilson M.C. and Gallup G. (1955). Extension Teaching Methods. Extension Service Circular 495, U.S.Department of Agriculture, Washington D.C.

Young V.P. (1998). Scientific Social Survey and Research Prentice Hall of India Pvt. Limited, New Delhi.